T P M

TOTAL PRODUCTIVE MAINTENANCE

Yoshikazu Takahashi
Takashi Osada

ASIAN PRODUCTIVITY ORGANIZATION

Published in 1990 by Asian Productivity Organization
4–14 Akasaka 8-chome
Minato-ku, Tokyo 107
Japan

Distributed in North America, the United Kingdom, and Western
Europe by Quality Resources
    A Division of The Kraus Organization Limited
    One Water Street
    White Plains, NY 10601

Originally published in Japanese under the title *TPM—Zenin
Sankano Setsubishiko Manejimento* by Nikkan Kogyo Shimbun, Ltd.
(The Daily Industrial News), Tokyo, Japan.

This is an authorized English translation from the Japanese
edition under agreement with the Daily Industrial News.

ISBN: 92-833-1109-4 (cloth)
     92-833-1110-8 (paper)

Printed in Hong Kong by Nordica International Ltd.
First printing 1990

Contents

Glossary of Acronyms and Abbreviations   vii

Preface   ix

**1   Advocacy and Importance of Equipment-Oriented
Management**   **1**

   1.1   Equipment is the Major Means of Production in
      a Factory   1
   1.2   The New Industrial Revolution: Technological
      Innovation in the 1980s and the Value of PM   2
   1.3   Factory Management During Slow Growth Periods
      and Maximum Utilization of Existing Equipment   5
   1.4   The Need for and Managerial Implications of a
      New Labor Structure Providing Incentive Work
      Environments—Role of TPM Activities   8
   1.5   The Toyota Production Method for the Machine Pro-
      cessing and Assembly Industry and the Need for TPM   14
   1.6   Perfect Production and PM Activities in the Integrated
      Petrochemical Industry Complexes and in the
      Steel Industry   16

**2   Setup of the Basic TPM Plan and Implementation by
Each Department**   **21**

   2.1   TPM Requirements   21
   2.2   The Sphere of Activities for Improvement of Overall
      Equipment Efficiency   22

2.3   Relationship Between Productive Maintenance and
      Plant Management Methods   23
2.4   Total Productive Maintenance Goals and the Process
      of Implementing Productive Maintenance Department
      Goals   31
2.5   The Organization's Role in Advancing the Use of TPM   34

**3   Analysis and Diagnosis of Management Standards for PM
and Methods for Improvement   43**

3.1   Analysis of Current Realities of PM Activities
      and Management Standards   43
3.2   Analysis of the Relationship Between Production
      (P: Production Volume) and Equipment   44
3.3   Analysis of the Relationship Between Quality (Q)
      and Equipment   85
3.4   Analysis of the Relationship Between Cost (C)
      and Equipment   105
3.5   Analysis of the Relationship Between Delivery
      (D: Delivery Date) and Equipment   113
3.6   Analysis of the Relationship Between Equipment
      and Safety, Environment, and Pollution (SEP) Factors   115
3.7   Relationship Between Morale (M: Motivation)
      and Equipment   122

4   Planning and Managing Maintenance   147

4.1   A System of Productive Maintenance and Main-
      tenance Department Tasks   147
4.2   Key Considerations in Designing the Main-
      tenance System   152
4.3   Analysis of Equipment Efficiency and Development of
      a Plan for Improvement   157
4.4   Planned and Chance Maintenance   165
4.5   Diagnostic Techniques and Predictive Maintenance   189
4.6   Improvement of Maintainability   201
4.7   Spare Parts Management   215

5   Self-Initiated Maintenance   227

5.1   Background of Self-Initiated Maintenance   227
5.2   High Priority Activities in Self-Initiated Maintenance   229
5.3   Daily Inspection   237
5.4   Skill Training for Self-Initiated Maintenance   242
5.5   Steps to Be Taken for Self-Initiated Maintenance   244

**6 Maintenance Prevention Design and Equipment
    Engineering Technology**   **247**

  6.1 Importance of Equipment Engineering Technology   247
  6.2 Themes for Equipment Engineering   259
  6.3 System for Implementing an Equipment Engineer-
       ing Project   260
  6.4 Key Points in Designing for Cost Down   264
  6.5 Key Points in Designing for Maintenance Prevention   267
  6.6 Project Summary and Evaluation   272
  6.7 Initial Material Flow and MP Information Management   272
  6.8 Building Know-How and Technological Training   278
  6.9 Organization of Engineering Tasks   281

7 Equipment-Knowledgeable Personnel—Skills Training   **285**

  7.1 Engineering and Skills Needed for Equipment   285
  7.2 Training Programs   287
  7.3 Promoting Skill Training   292
  7.4 "Handcrafted" Maintenance Skill Training Program   295

**8 Effects of Preventive Maintenance and Maintenance
    Evaluation**   **299**

  8.1 Significance of Measuring the Effects   299
  8.2 Enhancing Effectiveness through the PM Rally   301
  8.3 Choosing Items for Measurement and Evaluation of
       Impact and Effect   309

  Bibliography   312

  Index   313

Glossary of Acronyms and Abbreviations

ABNF	Abnormal but nonfailure
AC	Acquisition cost
CAD	Computer-aided design
CAE	Computer-aided engineering
CAM	Computer-aided manufacturing
CC	Cost control
CCA	Critical component analysis
CD	Cost down
CP	Process capability
FTA	Fault tree analysis
IC	Initial cost
IC	Integrated circuit
IE	Industrial engineering
FMEA	Failure mode and effects analysis
FMS	Flexible manufacturing system
JIT	Just in time
LCC	Life cycle cost
LCP	Life cycle profit
LSI	Large-scale integration
MC	Machining center
MC	Maintenance cost
MP	Maintenance prevention
M-Q	Machine quality
MTBF	Mean time between failures
MTBQF	Mean time between quality failures
MTN-Q	Maintenance quality

MTP	Management training program
MTTR	Mean time to repair
NC	Numerical control
OC	Operation cost
OSI	On-stream inspection
OSR	On-stream repair
PC	Production control
PERT	Program evaluation and review technique
PM	Preventive maintenance
PM	Productive maintenance
Pro-MTN	Programmed maintenance
PSE	Plant security evaluation
QC	Quality control
QCC	Quality control circle
QL	Qualty assurance production line
RC	Running cost
SC	Sustaining cost
SEP	Safety, environment, and pollution
TPM	Total productive maintenance
TWI	Training within industry
UNKS	Unknown factors
VCR	Video cassette recorder
VE	Value engineering
WIP	Work in progress
ZD	Zero defect

PREFACE

This preface would be incomplete if I failed to introduce the late Dr. Yoshikazu Takahashi to the readers of this book. Dr. Takahashi was one of the founders of the Japan Plant Maintenance Association. He was one of the pioneers who introduced and established Total Productive Maintenance (TPM) in Japan, transforming conventional maintenance practices into *a science of management* specialized for equipment maintenance. Using his long-term experience as a management consultant, he also created specific analysis, diagnosis, and developmental systems.

Three years have passed since Dr. Takahashi first conceived of this book. His previous work, *A Productive Maintenance Promotional Manual,* was published by the Japan Plant Maintenance Association. It enjoyed high recognition both at home and abroad, and in the ten years since its publication has become part of the standard collection of a considerable number of libraries in Europe as well as Japan. Meanwhile, as TPM became part of many industries nationwide, it has undergone a series of significant changes with regard to its contents, characteristics, and levels of sophistication. Hence the need to produce a new book on TPM.

Dr. Takahashi succumbed to cancer in 1984, his wish unfulfilled. It is no exaggeration to say that his life was devoted to studies of Productive Maintenance (PM), and he never stopped asking the question: "What is PM?"

One of the tasks that Dr. Takahashi left with me is that of completing this book. He was my teacher during the ten years I worked with him at the Productive Maintenance Technology Laboratory, and I have always taken it as my mission to publicize Dr. Takahashi's teachings by working on this manuscript, which he could not complete, and taking on the task of completing it myself. When this book reaches the public, one of my missions will have been fulfilled. If, by reading this book, more people understand PM, Dr. Takahashi's hope will also have been fulfilled.

Today's productive maintenance has moved from conventional PM, centered upon maintenance, to TPM, which emphasizes total participation and the role of manufacturing operators. Current TPM activities are expanding further by including the participation of an entire company's personnel—technicians as well as office workers, whose activities would often develop into group work, involving cooperation among outside businesses, including subcontractors and other affiliated companies. TPM activities provide not only PM techniques but also an understanding of PM philosophy and its application to the improvement of work environments. They differ from typical company social events or activities organized by limited numbers of people. New TPM activities involve all company members, and their revolutionary impact begins with changes in the factory plants, in human consciousness, and in work environments to reform the company's structure so as to produce higher profits.

TPM activities should be considered as consisting of a whole set of PM activities that address each individual stage of the equipment life cycle. These activities involve such stages as study of the equipment, decisions about equipment specifications and design, manufacturing, installation, test operations, actual operations, maintenance, updating, and obsolescence. Considering the breadth of scope and depth of coverage, PM activities present a staggering number of difficult problems, such as the need for classifying and organizing subject matter, defining practical implementation procedures, and educating the people who carry them out. These tasks are challenging and rewarding, and the more one becomes involved, the closer one approaches the core of the problems.

This study of PM philosophy is intended as a reference manual for use by TPM promoters as well as factory managers.

Chapter 1 focuses on the concept of "Equipment-Oriented Management," a philosophy also expounded by the late Dr. Takahashi. This chapter deals with ways of increasing the productivity level through total employee participation, without sacrificing respect for the human dimension, against a background of the ongoing transition in production means from human labor to machinery. Examples are taken from the production methods used at Toyota Motors Corporation and other processing industries. Chapter 2 is about TPM application and extension, the most vital step in the actual implementation of TPM.

Chapter 3 is concerned with analyses and diagnoses of the relationship between PM and such production considerations as productivity, quality, cost, safety, and motivation: TPM specifically emphasizes these five elements.

TPM aims at creating highly efficient production lines through maximum use of existing equipment with "zero breakdowns." Methods of applying TPM to Quality Control (QC) are currently under study; however, we have sought to establish here the concept of quality maintenance. It calls for the creation of a Quality Assurance Line that may ensure 100 percent reliable quality products. Plant enhancement and mechanization are also described in accordance with the idea of safety management. The concepts of the five management strategies (5S's) for increasing motivation through TPM activities now stand on their own. TPM motivation is based on these five strategies as well as on self-maintenance, the goal of which is the creation of a "disciplined work environment" where workers will learn about and take pride in the machines around them.

Chapter 4 covers maintenance planning and management, the foundations of productive maintenance philosophy. Chapter 5 deals with "Self-Maintenance" that, together with "Technical Education" (described in chapter 7), seeks to educate the people who will be thoroughly acquainted with machinery. People change through PM activities as they acquire and exercise new techniques and see themselves improving and receiving positive feedback. Heightened self-esteem and increased work satisfaction are often experienced by operators who are involved in this process of improvement.

Chapter 6 introduces a new type of plant construction, namely the building of functionally unique and distinctive plants that are satisfying to the people involved. Unlike passive conventional Maintenance Prevention (MP) design-oriented activities that only reflected reactions to past problems, new MP engineering activities are proactive, with a focus on equipment designs that are responsive to the unique requirements of each plant. Chapter 7 discusses technical education, and Chapter 8 is concerned with the achievement and evaluation of PM activities, treating achievement as a basis for further improvement.

Some parts of this study may be difficult to understand because of a lack of examples. The book, however, presents many new PM concepts, and I am confident that their implementation will prove useful.

This book was made possible by the cooperation of everyone who assisted in promoting TPM, those who aided in finding references, and the encouragement provided by Nikkan Kogyo Shimbun, Inc. I am grateful to all who helped me.

Lastly, I would like to dedicate this book to the memory of Dr. Takahashi. May his soul rest in peace.

Takashi Osada

1

Advocacy and Importance of Equipment-Oriented Management

Total productive maintenance—productive maintenance activities with total employee involvement throughout the enterprise—is among the most effective methods for transforming a factory into an operation with an equipment-oriented management that is consistent with the changes in contemporary society.

The first requirement for this transformation is that everyone (including the top management and managerial supervisors as well as operators), direct their attention to all components of the plant—molds, fixtures, tools, industrial instruments, and sensors—recognizing the importance and value of equipment-oriented management consistent with contemporary trends. Understanding equipment-oriented management is crucial because the reliability, security, maintenance, and operational characteristics of the plant constitute the decisive elements affecting product quality, quantity, and cost.

1.1 EQUIPMENT IS THE MAJOR MEANS OF PRODUCTION IN A FACTORY

A few years ago, figures published by the Japanese government's Bureau of Statistics indicated that Japan's gross national product had increased thirtyfold in the past twenty-five years while population growth was less than thirty percent. This growth resulted from the rapid replacement of manual manufacturing operations by mechanization via technological innovations, following the introduction of high-

1

speed, large-scale, continuously running, automated, complex machines with increasingly miniaturized operating components.

Today, it is not an exaggeration to say that no one except an artist could make a living through manufacturing by hand. There is a wealth of industrially manufactured goods startling in their designs and complex details. But when the production processes are examined, we find young workers flawlessly assembling these goods in a wholly matter-of-fact manner.

Innovation has simplified manufacturing processes as well as improved production design and quality. Engineering innovations in equipment have lowered the level of skill required for those operations still done by manual labor. Furthermore, as machines and equipment become more advanced, the number of parts also increases, making it difficult to achieve machine reliability and prevent breakdowns. For example, there are 10^2 parts in radios, 10^3 parts in televisions, 10^4 in automobiles, 10^5 in jet aircraft, and 10^6 in an Apollo spacecraft. If there are 500 parts and each one has a 99.99 percent reliability per unit of time, combined reliability is reduced to as little as 96.24 percent. It is therefore imperative to assure not only that parts are designed to be reliable but also that preventive maintenance methods are more precisely calibrated and well proven. It is generally accepted that personnel required for maintenance are 1.5 times higher when production capacity increases 100 percent. Thus, as mechanization progresses, more maintenance personnel and support will be required.

1.2 THE NEW INDUSTRIAL REVOLUTION: TECHNOLOGICAL INNOVATION IN THE 1980S AND THE VALUE OF PM

Technological innovation in the 1980s is said to have begun with electronics. Despite the reductions in plant investment, number of employees, and input energy caused by the first oil crisis, the Japanese GNP has grown dramatically because of plant technology innovation and the establishment of total participation production, maintenance, and quality assurance technologies. Japan's exhaust restriction is the strictest in the world, and it helped make big progress in promoting antipollution technology. The oil crisis also contributed to the progress of energy conservation technology. Japan's plant export is competitive because of these two technologies.

Especially with regard to the automobile, steel, and cement industries, Japanese energy conservation technology may be the most advanced in the world. In the 1970s, moreover, LSI (large-scale integration) appeared to replace IC (integrated circuits). The VCR (video cassette recorder), which was developed as a vanguard application product of LSI, has taken over almost the entire world market, and today is a $2 billion to $3 billion industry. The VCR's value as an educational tool is appreciated even more in developing countries than in developed nations. Development and marketing of the VCR is only the beginning. Astute businesspeople are already launching new LSI application products in a variety of industries.

This phenomenon is also influencing PM activities. Only a decade ago, computers were too expensive to use for controlling machine tools. Today, Japan is producing the largest number of numerically controlled (NC) machine tools in the world. Approximately one-half of the machine tools manufactured in Japan are equipped with NCs, whereas the number of NC tools in the United States is one-half of that in Japan, and is even lower in West Germany. In Japan, sixty to seventy percent of NCs are being used in small and medium enterprises. Forty percent of these industries, with fifty to ninety employees, have introduced some type of NC-based fabrication machinery. It will not be long before NCs will be used in every manufacturing company.

Furthermore, a new type of numerically controlled multimachine work station called a machining center (MC), which performs different tasks automatically with multiple machines, is also rapidly increasing in number. And even NCs and MCs may fade from the industry in the near future, because programming technology has become simpler than in the past. It has become increasingly possible to develop one's own programs, and even medium and small enterprises have begun in-house programming. The high percentage of NCs used in medium and small enterprises is considered unusual outside of Japan.

When NCs are combined with part-dislodging robots, multiprocess manufacturing can operate nonstop, even overnight. At present, this type of operation can enable an investment to be recovered within two years. Consequently, small- and medium-sized enterprises are enthusiastic about the application of this combination.

Following the utilization of robots, a flexible manufacturing system (FMS) will be used. FMS links a machining center to various types of robots, such as material conveyance and staging and dislodging robots. In this highly flexible system, it is possible to achieve automated manufacture of a variety of products in small quantities while drastically reducing the required number of manufacturing operators.

The revolutionary nature of large-scale integration's commercial value is rapidly being recognized in a variety of fields. The largest LSI user may well be the automotive industry, followed by the agricultural and construction machinery industries. Reflecting this trend, some Japanese products are rapidly gaining an upper hand in international competition, as exemplified by the recent development of the world's largest dump trucks and large farm machines, in addition to the traditionally competitive small- and medium-sized agricultural equipment. It appears almost certain that the technological innovations brought about by LSI will become crystallized in new and prominent Japanese industries. Moreover, this trend will continue, although the speed with which it will transform other Japanese industries may depend on foreign trade factors.

Not only are technological advances greatly affecting the machinery of plants, but the process of manufacturing is also undergoing rapid change. This poses a challenge with regard to the complete implementation of unmanned, automated operations and the transformation of manufacturing operations into continuous flow manufacturing. In machine processing and assembly industries it is difficult to assure reliability of a series of plant units, especially when dealing with hardware that assumes different shapes as its models change. Nonstop operation in this group of industries is even more difficult than in process industries, even though we are dealing with the same equipment-centered industrial innovations.

Unlike process industries, where the materials being processed are gases, liquids, powders, pellets, and the like, the objects of machine manufacturing and assembly are solids of various kinds and shapes that change geometrically together with model changes. To cope with the complexities arising from such differences, many problems arise with regard to engineering definitions and the setting of standards for procedures that might have been taken for granted in the past. These include item lineup, positioning, line balancing, cleaning, and parts ejection and dislodging. Unless these areas are improved, the implementation of unattended nonstop operation will never succeed. Thus, of major importance are the processes that lead to unattended operation and elimination of the need for manual machine adjustments.

Unlike people, NCs and robots provide steady quality control. Since the time required to complete a work process is readily determinable, delivery times may be strictly adhered to. Moreover, since they allow nighttime unattended operation, it is also possible to compress before-delivery lead times. Unlike human workers, machines will never be absent from work or require labor management. However, ideal as this may seem, do machines really work that well? Even jet aircraft manufacturing that boasts of its reliability is suffering from too many parts and from too high a ratio of parts breakdowns and mean time between failures (MTBF).

Machines deteriorate over time unless they are properly maintained. The duration of parts varies, depending on the amount of stress from operation and environments. Moreover, equipment precision affects product quality, and low precision can cause quality deteriorations. If a large volume of scrap is produced before anyone notices the problem, not only are the expected profits lost but the costs of the raw materials themselves may be squandered. Although sensors can be developed and used to detect inferior products, they may also reduce productivity by causing frequent brief stoppages and thus disturbing the processing flow.

How should all these problems be addressed and resolved? We must establish a maintenance technology that clarifies the causes of machine breakdowns and inferior products, based on analyses of the modes of equipment malfunctions and the types of resulting defects. These problems should also be approached from the perspective of equipment-oriented management, including total employee participation. The goals are to increase the reliability of equipment with regard to its quality of output products, avoidance of breakdowns, maintainability, safety, and operability as a program for promoting on-site engineering know-how. This precisely describes the goals of TPM activities. The modern competitive environment leaves us no other choice (see Table 1.1).

1.3 FACTORY MANAGEMENT DURING SLOW GROWTH PERIODS AND MAXIMUM UTILIZATION OF EXISTING EQUIPMENT

Japanese industries, which lead technological innovation in a number of fields, are expected to make tremendous progress in the long range. International economic influences have been so strong, however, that we may encounter difficulties that prevent short-term progress. To cope with such ever-changing circumstances, new types of industrial structures need to be built. Last year, the president of Toyota Motors described the necessity of creating a new company structure that would enable them to survive at a seventy percent operation rate, including the operation of all subsidiaries. With everyone's help, he succeeded in raising profits. It is clear in this example that lowering the break-even point to the limit is the key to creating a flexible industrial structure.

TABLE 1.1 Total Technology Required for PM Activities

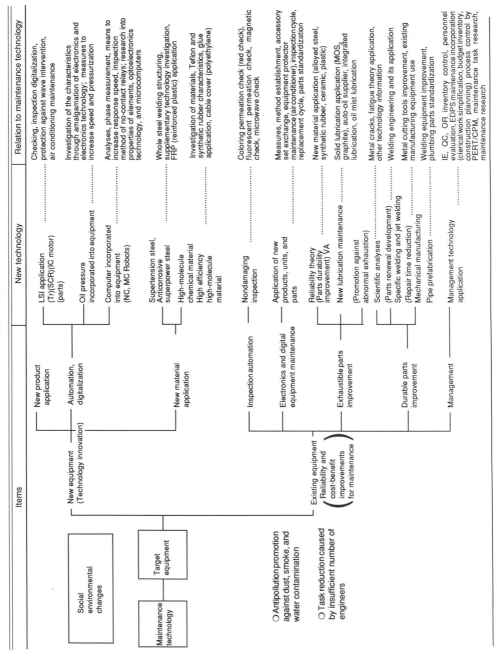

Items		New technology	Relation to maintenance technology
New equipment (Technology innovation)	New product application	LSI application (Tr)(SCR)(IC motor) (parts)	Checking, inspection digitalization, protection against wave intervention, air conditioning maintenance
	Automation, digitalization	Oil pressure incorporated into equipment	Investigation of the characteristics through amalgamation of electronics and electronics technology; measures to increase speed and pressurization
		Computer incorporated into equipment (NC, MC Robots)	Analyses, phase measurement, means to increase response speed, inspection method of no-contact relays; research into properties of elements, optoelectronics technology, and microcomputers
	New material application	Supertension steel, Anticorrosive superpower steel	Whole steel welding structuring, supplementary technology investigation, FRP (reinforced plastic) application
		High-molecule chemical material High efficiency high-molecule material	Investigation of materials, Teflon and synthetic rubber characteristics, glue application, cable cover (polyethylene)
Existing equipment Reliability and cost-benefit improvements for maintenance	Inspection automation	Nondamaging inspection	Coloring permeation check (red check), fluorescent permeation check, magnetic check, microwave check
	Electronics and digital equipment maintenance	Application of new products, units, and parts	Measures, method establishment, accessory set exchange, equipment protector maintenance (air conditioner), inspection cycle, replacement cycle, parts standardization
	Exhaustible parts improvement	Reliability theory (Parts durability improvement) VA	New material application (alloyed steel, synthetic rubber, ceramic, plastic)
		(Promotion against abnormal exhaustion)	Solid lubrication application (MOS₂ graphite), auto-oil supplier, integrated lubrication, oil mist lubrication
		Scientific analyses	Metal cracks, fatigue theory application, other technology information
	Durable parts improvement	(Parts renewal development) Specific welding and jet welding	Welding engineering and its application
		(Repair time reduction) Mechanical manufacturing	Metal cutting tools improvement, existing manufacturing equipment use
		Pipe prefabrication	Welding equipment improvement, plumbing parts standardization
	Management	Management technology application	IE, QC, OR inventory control, personnel evaluation, EDPS maintenance incorporation (clerical work simplification, budget inventory, construction planning) process control by PERT/CPM, maintenance task research, maintenance research

Social environmental changes

Target equipment

Maintenance technology

○ Antipollution promotion against dust, smoke, and water contamination

○ Task reduction caused by insufficient number of engineers

Innovative technology must be aggressively implemented in manufacturing operations. At the same time, the following points must be remembered: (1) restrict unnecessary equipment investment; (2) make maximum use of existing equipment; (3) enhance rate of equipment utilization for output; (4) ensure product quality through the use of equipment; (5) perform low-cost personnel reduction through equipment enhancements; (6) reduce the costs of energy and source materials through innovations in equipment and improvements in the methods of their use. All these tasks are fundamental to reforming a company structure, in response to the challenges of the 1980s and 1990s. They must be carried out with total employee participation.

Thus, TPM is a corporate-wide campaign involving the entire body of company employees, to challenge the full use of existing equipment to its maximum limit, using the equipment-oriented management philosophy. TPM activities include:

1. Investigating and improving machinery, molds, and fixtures that are reliable, maintainable, and safe, and exploring how to standardize these techniques.
2. Determining how to provide and guarantee reliable product quality through the use of existing machinery, molds, and fixtures, and how to train all personnel in these techniques.
3. Learning how to improve operation efficiency, and how to maximize operation duration.
4. Discovering how to educate operators so that they take an interest in and care about plant machines.

Two TPM concepts are especially effective for industries:

1. The development of new products, processing techniques, or machine technology is more likely to be done by a small number of competent engineers. This is especially true of scientific advancement that has always been led by a few able scientists or engineers. Industries have also progressed as a result of improvement in products or machine technology that was promoted by a few industrial engineers.
2. In production activities, however, nearly all the employees of a company are engaged in sustaining production by use of machines, molds, fixtures, and tools, guaranteeing product quality and timely delivery, and promoting reduction of work-in-process inventory. Quality improvement of these workers does not take place over a short period of time. Consequently, commitment by an individual

worker is required to promote reduction of costs and the number of inferior products and to guarantee timely delivery.

TPM focuses its management points on machinery, molds, and fixtures, and highlights the following:

- Determining product quality by means of appropriate equipment.
- Production and delivery controls through appropriate equipment.
- Assurance of environmental protection and safety through equipment management.
- Education of workers by interesting them in the machinery, molds, and fixtures with which they are working, so as to internalize a sense of respect for hardware. On a long-term basis, an infrastructure can thus be developed within corporate human resources that is thoroughly familiar with the machinery, regardless of its type.

Since such employee nurturing is a difficult and painstaking task, managers and administrators must have both a very strong sense of commitment and a long-term perspective. Many workers, if trained through equipment-oriented management with a five- to ten-year time perspective, would prove their worth not only through their contribution to improved company productivity, but they would also provide new corporate "muscle" in competition with other firms. Unlike externally purchased technologies, the sharpening of the workers' skills thus developed from the grass roots up and the concomitant strengths of the corporate culture are not easily achieved. This is why PM, based on a long-term perspective, is considered one of the effective methods of transforming the qualitative contents of a plant.

1.4 THE NEED FOR AND MANAGERIAL IMPLICATIONS OF A NEW LABOR STRUCTURE PROVIDING INCENTIVE WORK ENVIRONMENTS—ROLE OF TPM ACTIVITIES

1.4.1 Need for and Value of Division of Labor

The need for diversification of skills and specialization, and the value of such specialization, was first advocated about fifteen years ago, in business literature and seminars. Even earlier, soon after World War II, the concept of division of labor and specialization was prevalent in Japanese industries, making all the difference in their scope and variety. Japanese industries were then struggling to recover from the effects of the war and were aggressively searching for ways to introduce plant technology, change manufacturing methods, and create new types of organizations. These circumstances encouraged acceptance of the concept of labor specialization. Labor separation clarifies task responsibility and helps organize manufacturing structures, allowing tasks to be learned in a short period of time.

The idea of separating tasks according to type is said to have come from the American employment system and the structure of the U.S. Army, Navy, and Air Force. In the American employment system, individual responsibility is more clearly defined than in Japan. Americans are well informed about their individual abilities, and they often work on improving themselves, in order to seek promotion and better-paying jobs. Moreover, American labor unions are separated according to type of work, and the distinction between jobs is so specific that one is not welcome to work in a job classification other than his or her own. Even in the United States, however, labor separation (specialization) is not dominant, and a conflict exists there between specialization and generalization. The American employment system is considerably different from that of the Japanese, which assumes lifetime employment.

Military organizations consist of many people who had been engaged in various kinds of work before joining the armed services. When one looks at how the modern army is organized, one is amazed at how much more emphasis is placed on maintenance of arms than on the battlefront. In the army, specialization is necessary for rapid education of maintenance personnel.

The Japanese educational system owes much to the model provided by the U.S. Army, and the concept of lifetime employment was called into question during the post-war period. In this context, division of labor in Japan reached a significant level of development.

1.4.2 The Need for and Significance of Broadening the Scope of Job Assignments—Change in the Concept of Craftsmanship

Specialization is a labor system that provides clear distinctions in responsibility and short-term task acquisition. Although it must be applied carefully according to needs, it is highly useful in the fields of research and development, as well as in individual and management technologies. In industries that are dependent on mechanized facilities, however, areas of technical skills can be seen as gradually expanding. Compared to ten years ago, more and more industries are taking a serious interest in broadening the scope of technical skills.

Too great an emphasis on specialization has resulted in workers lacking responsibility, having insufficient understanding of overall production systems, and having excessive idle time due to personnel changes. Consequently, current emphasis on broadening the scope of skill requirements tends to stress improvement in these areas. The real value of task generalization lies, however, in creating an incentive work environment based on human respect.

The more rapidly technological innovation occurs, the faster the obsolescence of technical skills progresses, shortening the available life of acquired techniques. To eliminate idle time caused by the installation or addition of plant equipment, workers are urged to acquire skills in a short period. Both phenomena have had negative effects on the experienced worker and have lowered the quality of workers' skills.

In the Japanese lifetime employment system, people are promoted according to age. This system may cause problems. For example, a newly assigned administrator over the age of forty-five cannot successfully supervise high school graduates, because his past experiences are not applicable to the operation of new machines and he lacks knowledge about new technology. Production-line workers who repeat the same type of tasks may lose their motivation, perceiving that they could find more happiness in leisure rather than in doing the same thing for ten years. Thus, satisfying work environments can be created not by merely discussing "motivation," but by providing workers with tasks that give them an incentive to work.

The Japanese lifetime employment system must be adapted to cultivate individual talents, enabling them to be used effectively for the company's activities. Some employees may regret having wasted their time in work because their talents were not utilized. They may be at fault for not having cultivated themselves, of course, but the company can also be blamed for its negligence in failing to further their workers' education.

The Japanese Ministry of Education reports an educational trend showing that more than ninety-four percent of entry level workers are high school graduates. To avoid wasting workers' talents, it is important to have a clear perception of what kinds of workers we are looking for and what sorts of talents we should cultivate. As noted before, mechanization is now replacing manual operation by experienced workers. Thus for all types of tasks, PM techniques are more necessary than ever to ensure quality, timely delivery, low cost, safety, and pollution prevention. Veteran workers should be released and exempted from such tasks as switching, packing, and visual inspection of products.

Traditional, highly skilled Japanese carpenters are known to take special care of their tools, continually trying to improve their functioning and efficiency. The same attitude can be applied to mechanical manufacturing. Since the Industrial Revolution, machines have evolved from hand tools. To be able to make full use of them, people who work with machines should be familiar with their structures, functions, and even with their defects. Going further back in history, it has been suggested that the Roman Empire collapsed because the Romans gave away their advanced manufacturing technology to the Germanic peoples, whose technology outpaced that of the Romans by the time the empire fell. In the Roman Empire, manufacturing was done primarily by slaves. In ancient Germany, it was done by free artisans who were zealous workers, and who maintained and improved their technology far more than the Romans had.

A contemporary example is afforded by veteran auto racers, who call their race cars "machines." Before each race, their machine's performance and structure are carefully studied, and machine conditions are monitored during long-distance races. These professionals know that driving techniques alone will not win a gold medal unless those techniques are combined with good maintenance. Another example is that of someone aspiring to be a boiler operator in a large plant. He or she is aware of the need to pass a special state examination, covering such areas as functional structures and autocontrol circuits of highly advanced, complex power plants. To pass this examination, one must have a full grasp of the power plant system. Future maintenance thus requires proven techniques for carrying out plant regulation and troubleshooting.

As both quality control (QC) and zero defect (ZD) campaigns have proven, careless mistakes or inferior production due to human negligence can be avoided by maintenance activities.

For more complex problems, however, relative analyses must be performed, based on the relation between the plant and the manufacturing or chemical reaction processing technology. Some of these problems can be solved by studying source materials, catalyzers, and chemicals, but most will be resolved by improving machinery, molds, and fixtures.

For example, some QC campaigns have been successful because they modified plant machines, molds, and fixtures. Successful QC and ZD campaigns will increase the number of workers who are thoroughly acquainted with mechanical facilities. PM activities help provide better quality control, timely delivery, lower costs, and a reduced number of processes. These activities are the foundation of management as well as the core of productivity improvement. They must be implemented and well established before any new management procedure is introduced.

To promote PM activities, supervisors in all departments must first understand and support PM. The company president and heads of production, personnel, and labor education departments must do the same to further extend PM activities in other departments.

QC and ZD campaigns that have been promoted by the production department will not counteract PM activities but will reinforce them. Training within industry (TWI), management training programs (MTP), industrial engineering (IE) fundamentals, and QC guidance can all be adapted and strengthened to open up new horizons in technical skill training.

1.4.3 Technical Skills in the Workplace

Technical skills will not become obsolete. Young workers should be encouraged to learn a wide variety of techniques although it may take them a relatively long period of time to do so. It is first necessary to sort out the types and characteristics of technical skills required for factory activities. It is then possible to begin long-term educational planning to reorganize existing task areas according to individual aptitudes.

Basically, there are three types of skills in industry: (1) those that are useful for a long period, (2) skills requiring time to acquire and that are developed to help older workers establish leadership, and (3) those that cover general fields. These should all be integrated and modified so that they are consistent with individual task areas. Task area reorganization will help us to envision how novice workers will develop in ten to twenty years; such reorganization will ultimately help produce pleasant work environments.

For labor reassignment to be successful, task area reorganization should be initiated by taking a skill inventory among the workers of the plant. At the same time, it is necessary to evaluate and choose prospective skills that satisfy the three conditions just given. The reorganization should also include clarification of essential training goals, according to the needs at the current site, that can be assigned to workers as tasks. Another effective way to promote workers' involvement in their tasks is to give them challenging assignments through group activities.

PM activities demand tremendous efforts by shop floor supervisors as well as by the staff, personnel, and labor departments to produce better workers and working conditions. These activities are the prerequisites not only for increased productivity but also for maintaining older workers' leadership, for eliciting a level of service appropriate to individual wages, and for establishing the status of long-term workers.

It should be emphasized that the technical skills related to PM activities must meet the three conditions just mentioned. Technical skills include daily checking, equipment tuning, precision inspection, lubrication, adjustments, troubleshooting, and repair. These skills cover a broad range of task areas and should be taught to workers only after definition of a worker's individual task requirements. One must be careful not to make a hasty task allocation, which may result in inferior maintenance. Thus, workers should have a complete grasp of productive maintenance before any tasks are assigned. Task allocation must be consistent with the type of processing and "fit" task characteristics.

As automation progresses, maintenance technology areas will broaden, requiring speedy acquisition of new technology. Automated systems are, however, mainly positioned in the "brain" part of the plant, while the plant "body" very often consists of energy conducting, moving parts, and tank- or container-related piping. Even in the future, this structure will not easily change. Consequently, PM's technological skills will have long-term utility. Plant innovation is typically accomplished by a few competent engineers. On the other hand, it is no simple matter to develop maintenance personnel capable of using an innovation for performing production maintenance, quality control, timely delivery, and smooth operation. Thus, employee education must respect the talents of individual workers. TPM is an effective approach to achieving this type of education.

Task reorganization on a long-term basis cannot be accomplished solely by the plant-related sections. Their efforts must be accompanied by educational planning and a reexamination of wage systems, the initiative for which must come from the personnel, labor, and education departments. Total participation is thus mandatory for cultivating talent over the long term and for creating stimulating work environments.

1.5 THE TOYOTA PRODUCTION METHOD FOR THE MACHINE PROCESSING AND ASSEMBLY INDUSTRY AND THE NEED FOR TPM

To systematize all the lines of automobile manufacturing into a Just-In-Time (JIT) system is a challenge for the machine processing and assembly industry. It is a challenge to facility mechanization which contradicts the nature of machine processing and assembly manufacturing; however, results from this type of systematization are remarkable, especially with regard to cost reduction. PM activities are necessary to sustain systematization.

The former president of Toyota Motors, Kiichiro Toyoda, stated this principle in his book, *Action Then Theory*. He was convinced that, in parts-assembly industries such as Toyota Motors and in assembly manufacturing in general, it is best for parts to be assembled on the line "just-in-time." He applied this conviction for thirty years, overcoming obstacles one by one, until formulating this principle. His theory of processing maintains that one takes what is necessary when it is necessary. This means that each assembly line produces as much as is required for the following line. To communicate a message about a part, an operator attaches a simple task-direction note called a "kanban" to each piece to be processed, which is placed in a plastic bag.

This apparently insignificant procedure has had a major effect. Toyota assembly lines handle a total of more than 20,000 automobile parts through all operational steps. This simple procedure aims at reducing the work-in-process (WIP) inventory parts to zero. It suggests that parts flow during each of the manufacturing operations as if they were water flowing in a pipeline.

The flow of parts typically stagnates because of the following:

1. Variations in the production capabilities of different work centers build up WIP inventory.
2. When the staging or setup times for parts changeovers are long (because of frequency, setup time, or complexities of operations), either the WIP inventory of the parts increases or the subsequent lines come to a halt.
3. When the load balance of two consecutive operations are not matched, either the WIP inventory of their parts increases or the subsequent operations come to a halt.
4. When the number of quality defects is high in one operation, either subsequent operations cease or there arises a need for buffer inventory of the parts in question.

5. When equipment breakdown occurrences are frequent, or there is a high rate of repair time/frequency, subsequent operations will stagnate.
6. Frequent personnel absences or poor personnel management can result in direct negative effects.
7. Unsafe equipment or behavior can have a significant negative effect on the entire production operation.
8. Other circumstances.

In an effort to solve such problems, Toyoda applied the production methods of a processing industry to mechanical assembly. However similar these two production methods may appear on the surface, they are substantially different. Thus, one can appreciate the tenacious efforts required to develop better production methods.

With regard to oil refineries and the petrochemical, chemical, and paper industries, maintenance technology and management have been developed, since these were crucial needs in those process industries. Although the type of processing varies, depending upon whether the source material is a gas, liquid, or solid, capacities between lines are by and large well balanced in the process industries. The flow between pipes or conveyors is usually smooth. There are a few intermediate tanks between different processes, but these are used to control mixture unevenness and quality, rather than for product storage. Their WIP surplus is only worth a few hours, and the number of such intermediate storage locations is minimal. Consequently, when a processing flow stagnates because of a machine breakdown, the loss is great. If there are fewer backup machines, it is more difficult to sustain overall processing reliability. Maintenance management is therefore indispensable to avoid fatal consequences. Toyoda recognized the applicability of this experience to developing more profitable manufacturing conditions in his factory.

Even today, vertically integrated petrochemical industries have been investigating a desirable method of maintenance management with an ultimate goal of operating nonstop 330 days per year under zero breakdown and zero accident conditions. In the Toyota production method, automobile production lines are linked through the "Kanban System" described before (a system of material and production control maintained by attaching a simple card to the parts bin).

When the number of WIP inventory parts is minimized, movement of materials through their assembly lines will come close to that achieved in process industries, even if parts in some areas are conveyed via a hand-pushed cart. Because of the enormous number of parts dealt with on the Toyota assembly lines, it is difficult for every line to practice this minimization thoroughly. However, the result realized by the efforts of both workers and educators is greater than expected, and their improved management will continue to be an asset to their company.

The stress on productive maintenance activities generally increases or decreases in proportion to production demands: it decreases when the demands lower and increases when the demands rise. Toyota's maintenance activities, however, clearly establish the need for productive maintenance regardless of the proportion of production demands. Following the motto: "Make what is needed—as much as needed—when it is needed," Toyota's goal is a zero WIP inventory of parts. The Toyota production method refers to an equipment utilization ratio as an "operational ratio." This signifies that their equipment is ready to operate whenever necessary, and that after the equipment has been activated, there are no defect outputs. Toyota is trying to achieve the rough equivalent of the Apollo space program's goal: to eliminate any failure between the Apollo's departure from and return to earth after completing its mission to the moon.

Regardless of the ratio of production demands, the Toyota method asks production and maintenance technicians to ensure the planned life expectancy of the equipment, using minimum investment and maintenance costs. This may be asking a great deal, but it provides technicians with challenging goals.

The PM technicians must be careful not to put too great an emphasis on bought technology, and should perform continuing studies regarding productivity improvement and cost reduction.

Technicians can meet the challenges and solve many of these problems by planning process and equipment designs aimed at elimination of waste. The image of an ideal worker in the context of PM activities is one who can accurately perceive the manifestation of waste, and translate observations into specific changes in operational processes and equipment. The role of PM is indeed quite significant in both the Toyota production and productive maintenance methods.

1.6 PERFECT PRODUCTION AND PM ACTIVITIES IN THE INTEGRATED PETROCHEMICAL INDUSTRY COMPLEXES AND IN THE STEEL INDUSTRY

Production in integrated petrochemical industry complexes is usually carried out by interdependent factories linked to each other via pipes. A machine breakdown may affect a variety of operational processes, requiring a significant amount of time for recovery and adjustments not only in material but also in steam pressure balances. The resultant overall losses depend upon the specific situation, but, in Japan, they may reach a few hundred million yen per day. Thus these industries are challenged to achieve perfect production with absolutely no breakdowns. Those personnel dealing with this challenge cannot tolerate even two or three breakdowns per year. They must guarantee perfect operation, with only one shutdown for the annual repair, maintaining nonstop operation for 330 days per year. To achieve this goal, even the slightest negligence in shutdown checking, repair planning, or quality control is unacceptable.

The author once worked to establish the PM system with highly qualified production and maintenance technicians in an integrated petrochemical industry complex (Figure 1.1). Careful research and timely analysis of primary plant components and repair planning was carried out. In the course of periodic checking, we were all determined that such checking should guarantee nonstop operation for at least one year. I remember the workers' enthusiasm as they shared the conviction that each would perform a perfect job to bring about perfect production.

Accidents such as fire, explosion, and pollution at a giant production facility complex have a major impact not only on a company, but on society as a whole. PM activities, therefore, focus on these concerns as well.

One PM task is the creation of a totally innovative concept of production activities, requiring operators to achieve a much higher capability level than in the past. For example, such a task as monitoring plant conditions may appear to be simple; however, it involves foreseeing a slight defect in gigantic facilities and maintaining safe operation over a long period of time. These operations are highly complex and require a maximum interface between the human brain and computer. As facilities are mechanized, automated, and enlarged, operators are required to learn their functional principles and structures. They may need to possess the kind of skills and knowledge comparable to an airline pilot or flight engineer. I suggest that in the future a rigid system for a national qualifying examination be established and that a company not be allowed to operate unless it employs a certain minimum number of qualified workers who have passed such an examination. To achieve this goal, current personnel management systems need to be modified.

Maintenance technicians must perform thorough diagnoses of a plant that cannot be shut down for checking. They should also investigate how stress from the operation or environment is related to the

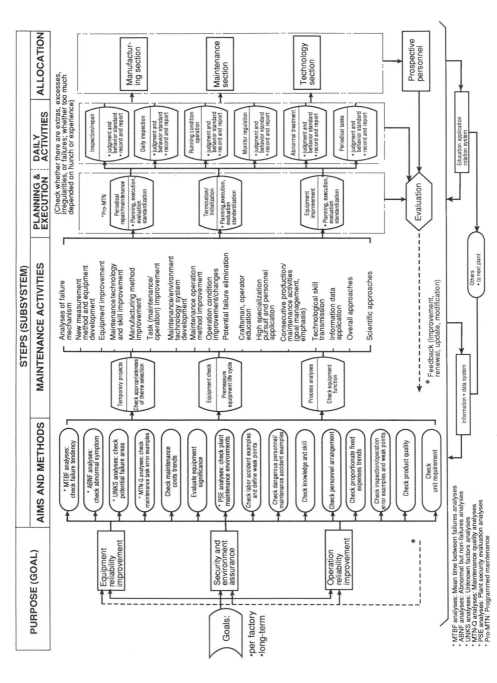

FIGURE 1.1 TPM Master Plan in the Petrochemical Industry (Model)

* MTBF analyses: Mean time between failures analyses
* ABNF analyses: Abnormal but non-failures analyses
* UNKS analyses: Unknown factors analyses
* MTN-Q analyses: Maintenance quality analyses
* PSE analyses: Plant security evaluation analyses
* Pro-MTN: Programmed maintenance

ratio of machine breakdowns, and how a system of diagnosis can be applied. The number of areas to be dealt with is unlimited, and the work is challenging, requiring highly advanced technical skills.

Japan is known for having the world's most productive steel industry relative to the vast amount of equipment investments. But the maintenance of equipment in Japan is made possible through PM activities. The steel industry in particular has always emphasized maintenance. For example, Nippon Steel developed a remarkable technology for checking equipment in accordance with their company's management principles. To avoid loss caused by machine breakdowns, the steel industry typically spends a large amount on annual maintenance. International conference statistics reported in 1988 that the Japanese steel industry spends seven to eight percent of its sales on maintenance, while non-Japanese steel industries spend ten percent. Thus TPM activities aim at perfect production while reducing maintenance costs.

2

Setup of the Basic TPM Plan and Implementation by Each Department

The key to the establishment and development of the basic TPM plan is ensuring the support of the plan's priorities and activities by the top management who drive it forward. The most important point is how well the top and middle managers recognize the necessity for and future value of TPM activities.

2.1 TPM REQUIREMENTS

We are actively working to disseminate and advance TPM activities in the Japanese industrial community. As mentioned in chapter 1, these PM activities mean not only preventive maintenance, but also a wider PM that focuses on the lifetime economic feasibility of equipment, molds, and jigs that play the major role in production. The requirements for this kind of PM are: (1) creating equipment with the greatest possible overall efficiency; (2) establishing a total PM that takes into account the entire life of the equipment; (3) maintaining motivation by utilizing independent small-group activities; (4) covering the planning, utilization, and maintenance of the equipment; and (5) companywide participation, ranging from top executives to individual workers.

These requirements are advocated because of the importance of the following procedures:

1. Making plant and equipment investments as cost-effective as possible.
2. Promoting engineers with aptitudes for working with the technical innovations involved in continuous assembly lines and the equipment-based processes, and promoting and retaining plant engineers who have a system-oriented grasp of life cycle efficiency and economic feasibility, and the ability to improve them.
3. Developing a large base of workers who can understand and deal with the plant's basic maintenance procedures and production operations (see Figure 2.1).

2.2 THE SPHERE OF ACTIVITIES FOR IMPROVEMENT OF OVERALL EQUIPMENT EFFICIENCY

There are numerous activities that can improve overall equipment efficiency throughout the life cycle, but in simple terms, it is a matter of increasing the ratio of the profits gained from the equipment vis-à-vis their overall cost. The overall cost of equipment can be shown as the sum of the initial capital investment plus the operating and maintenance costs of production. In other words, it is the life cycle cost. Moreover, the calculation of profits from the equipment is extremely difficult. This would be a simple calculation if the plant operated at the same level throughout its life cycle, but the level of operation changes with business fluctuations. In the case of current need investments, the basis for calculation is relatively clear, but with strategic investments (such as the expansion and improvement of equipment and new investments that accompany increased production), the problem of estimating demand is ever present, and management's control over timely decision making is often the most important factor.

The life cycle costing used by the United States Department of Defense is applied primarily to weapons, and in their case there is no direct profit from the products themselves. If the products meet the usage requirements, it is sufficient to evaluate the total expense, i.e., the life cycle cost. In the TPM measurement of overall efficiency, the problem of output that accompanies the projections is taken into account in the annual total hours of operation. If that output is low, then the sales department must provide feedback on the manufacture of alternative products through a modification of equipment. Excluding the estimating problems, the profits generated by the equipment can be expressed by an index relating the value added and the operation and loading rates. The value added is the income derived when the cost of materials and subcontracted processing has been subtracted from the revenues generated by the product.

The various research items for improvement of these elements are given in the boxes in Figure 2.2. The underlying principle here is that the development, selection, purchase, improvement, and utilization of equipment, machines, moldings, fixtures, electronic devices, and instruments should all be carried out with the utmost skill. Obviously, there are various techniques in which IE and QC approaches may be used at the conceptual level, but, in the end, we must concentrate on the technology for equipment, machines, moldings, and fixtures in order to realize these concepts. TPM relates to activities that deal with the improvement of overall equipment efficiency in terms of equipment-oriented management, and we must tackle problems from that point of view.

2.3 RELATIONSHIP BETWEEN PRODUCTIVE MAINTENANCE AND PLANT MANAGEMENT METHODS

After World War II, a variety of plant management methods were introduced into Japan both from Europe and America. Much has been written about these methods, which have significantly influenced many aspects of plant management. Today, as younger personnel seek to learn more about them, managers are often sidetracked by the technical aspects of these systems, thereby losing sight of the most direct path to solving immediate problems. At times, people in different departments within a plant are not aware of the intrinsically interdependent character of some of these technically refined management methods, such as quality control, Industrial Engineering (IE), Production Control (PC), PM, and Cost Control (CC). As a result, there is confusion about the areas of overlap, inconsistency in levels of effectiveness, and subsequent marginal results.

Consider, for example, a situation in which a group of people within a plant become involved in emphasizing quality control, pursuing a quality assurance program led by their own quality control groups. Their efforts, however, may not be sufficient. Further efforts need to be made to lower the number of operational steps and to promote the kind of industrial engineering technologies that can reduce labor requirements. This should be undertaken as part of the plant management's ongoing goals. Other crucial tasks include the maximum improvement of production control in order to meet delivery dates and the reduction of costs based on systematized cost control meth-

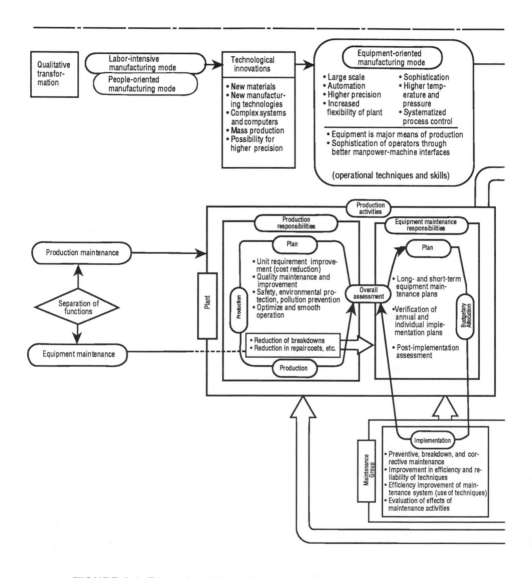

FIGURE 2.1 Example of Plant Management Based on Equipment-Oriented Management—Improvement in Overall Efficiency

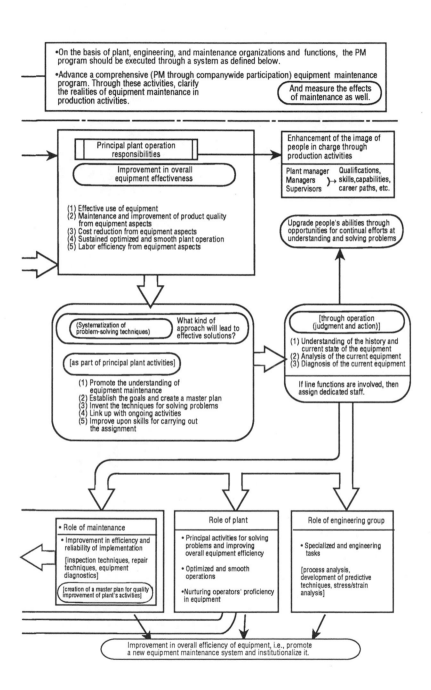

- On the basis of plant, engineering, and maintenance organizations and functions, the PM program should be executed through a system as defined below.
- Advance a comprehensive (PM through companywide participation) equipment maintenance program. Through these activities, clarify the realities of equipment maintenance in production activities.

And measure the effects of maintenance as well.

Principal plant operation responsibilities

Improvement in overall equipment effectiveness

(1) Effective use of equipment
(2) Maintenance and improvement of product quality from equipment aspects
(3) Cost reduction from equipment aspects
(4) Sustained optimized and smooth plant operation
(5) Labor efficiency from equipment aspects

Enhancement of the image of people in charge through production activities

Plant manager   Qualifications,
Managers   ⟩→ skills, capabilities,
Supervisors   career paths, etc.

Upgrade people's abilities through opportunities for continual efforts at understanding and solving problems

(Systematization of problem-solving techniques)

What kind of approach will lead to effective solutions?

[as part of principal plant activities]

(1) Promote the understanding of equipment maintenance
(2) Establish the goals and create a master plan
(3) Invent the techniques for solving problems
(4) Link up with ongoing activities
(5) Improve upon skills for carrying out the assignment

[through operation (judgment and action)]

(1) Understanding of the history and current state of the equipment
(2) Analysis of the current equipment
(3) Diagnosis of the current equipment

If line functions are involved, then assign dedicated staff.

- Role of maintenance

- Improvement in efficiency and reliability of implementation

[inspection techniques, repair techniques, equipment diagnostics]

[creation of a master plan for quality improvement of plant's activities]

Role of plant

- Principal activities for solving problems and improving overall equipment efficiency

- Optimized and smooth operations

- Nurturing operators' proficiency in equipment

Role of engineering group

- Specialized and engineering tasks

[process analysis, development of predictive techniques, stress/strain analysis]

Improvement in overall efficiency of equipment, i.e., promote a new equipment maintenance system and institutionalize it.

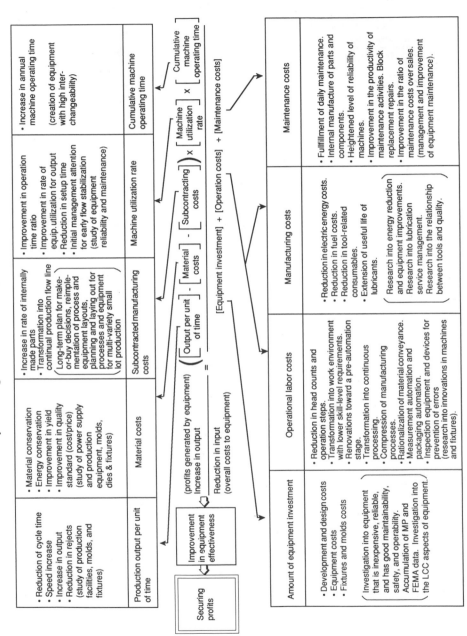

FIGURE 2.2 Various Elements for Raising the Overall Efficiency Level of Facilities

ods. Another goal, beginning with the design phase, is the lowering of production costs based on a Value Engineering (VE) approach.

In addition to the above, a program of productive maintenance (PM) should also be implemented. Productive maintenance is often regarded as too expensive. Justifications for this assumption include: (1) the low priority given to productive maintenance; (2) the lack of personnel available to carry it out successfully, despite the number of things that need to be done; (3) the feeling that too many meetings are held, with the result that an individual employee's efforts become too scattered and diffused to allow his or her undertakings to be completed with any significant results.

Various elements contribute to effective plant management, all of which are independent of the types of products being manufactured. These components incorporate other, more basic ones. It is the appropriate combination of these factors that needs to be considered when developing and implementing plant management goals and techniques. It is therefore relevant to clarify here the relationship between the methods used in plant management and a program of productive maintenance.

Table 2.1 shows the basic elements involved in plant management: personnel, machinery, and materials, all of which can be assigned a monetary value. These comprise the input for plant activities. Effective management methods are expressed in the skillful interweaving of these basic components. The combination of the 4M's—Manpower, Machine, Material, and Money—plus the M of Management Method, results in the concept of 5M Management. The output from plant management as it corresponds to the input may be defined as: P (Production volume), Q (quality), C (cost, unit requirement, work in progress), D (delivery date), S (safety, safeguards against environmental pollution), and M (morale, mental outlook toward work). These are directly related to the goals of the people on the plant management team.

Vital to management is the question of how to interrelate these six factors of output and the previously mentioned 5M's systematically, while adjusting to a specific business environment and establishing appropriate goals at any given time.

Let us analyze the interrelationship among management methods, input, and output. In Table 2.1, starting with row P (production quantity) and scanning to the right as the arrow indicates, we see a description of how to use varying numbers of workers and their skills effectively; which machines, facilities, and materials can be used and when they should be used—otherwise known as production planning and scheduling.

TABLE 2.1 Relationships Between Various Plant Management Methods and PM Activities—5 Elements of Plant Management (5M Management Method)

Output \ Input	Money			Method
	Man	Machine	Material	
(P) Production				Production Planning & Scheduling
(Q) Quality				Quality Control
(C) Cost				Cost Control
(D) Delivery				Delivery Control
(S) Safety				Safety & Pollution
(M) Morale				Human Relations
Method	Manpower Authorization	Productive Maintenance	Inventory Control	$\dfrac{Output}{Input} = $ Overall productivity Goals of plant activities

With regard to production planning and scheduling, a number of systematic methods have been developed and tested related to characteristics of the products manufactured and the production methods used. Some examples include the manufacturing cell—or subfactory-scheduling control techniques based on order dispatch lists, kanban scheduling control techniques based on visual control aids, and a computerized production scheduling system.

Similarly, as shown in Table 2.1, for each output factor, the management method is used to analyze the relationships among the input factors of personnel, machinery, and materials, and to relate each of these to effective plant activities. As outlined before, the management method incorporates quality control, cost control, delivery control, safety and pollution control, and human relations management.

In examining Table 2.1, as we scan the column downward from the input factor (machinery), we can see that productive maintenance (PM) is the element that links this input factor with each of the output factors to achieve effective results.

Specifically, productive maintenance deals with the following questions:

1. What kind of machinery and equipment planning is required to attain the desired production quantity?
2. Is there an appropriate balance between the intended used of various machines and their capacities?
3. Can the design characteristics of the machinery, tools, and other equipment meet the quality requirements of the product(s) to be manufactured?
4. How does wear and tear on the machinery relate to product reject rates, as the machines wear down after continued production use?
5. What is the relationship between losses and each of the machines? For example, to what extent are stoppage losses and decreases in yields caused by unexpected breakdowns? And what is the relationship between the unit requirement of electric power or other fuels and machine breakdowns?
6. How can costs for machinery maintenance be reduced, i.e., the costs of inspection, oiling, and repair?
7. How will machine breakdowns affect delivery time and the volume of orders to be released to the floor under any given production schedule?
8. What kind of relationships are there between machinery deterioration and aspects of safety and pollution control?
9. What should be considered in the relationship between the mental outlook of workers and automation, as manual production tasks become increasingly automated?
10. Does there exist an interest in the machinery and a sense of respect for its upkeep?
11. Are quality control groups promoting such attitudes toward the operation and maintenance of machinery, as well as suggesting improvements in machinery and ancillary components such as molds and fixtures?

Productive maintenance addresses and investigates all these questions, thereby promoting more effective plant activities.

Similarly, staff and materials management are also management methods to be studied in light of their relationship with the output from plant management, that is, P (production volume), Q (quality), C (cost), D (delivery date), S (safety), and M (morale).

The goal of plant management is always to increase the value of the index of the just mentioned input, or in other words, to raise the overall productivity level:

Output/Input = Overall Productivity

It is important to use appropriately the variety of management methods for enhancing overall productivity while at the same time focusing on plant operation cost reduction. Today, there is an ongoing general trend that is characterized by a shift from a mode of production that is heavily dependent upon skilled manual labor to one that is predominantly reliant on machines. Against this background, if productive maintenance methods are ignored, it is increasingly difficult to achieve quality assurance, delivery date adherence, cost reduction, and assurance of safety and environmental pollution prevention, regardless of which part of the operation is involved.

TPM focuses on these issues from the very beginning. This focus not only promotes the effectiveness of various plant activities, but can also have a marked effect on productivity improvement and cost reduction. There has been a significant increase in the number of plants where TPM has been introduced, with quiet but ongoing results.

As firms introduce new machines together with other technological innovations, emphasis tends to be placed upon acquiring the know-how needed to use these machines and to train the users of these new technologies. But of utmost importance is the nature of the culture that exists in the plant, with regard to accommodating and utilizing this technology. Central to the character of this culture is the number of seasoned workers who have experience with a variety of machines and technologies, experience that can only be gained through many years of nurturing.

In the future, it can be expected that personnel with a thorough proficiency in productive maintenance technologies will be needed. This in itself, moreover, may become the definition of a manufacturing worker. In the United States, for example, there have been times when General Motors took on the challenge of productivity improvement by actively adopting robotics. It was then reported that the surrounding setups were incomplete; workers in manufacturing were ill-prepared to respond to various kinds of problems, both major and minor; and maintenance engineers could not cope with all the demands expected, resulting in situations in which production engineering staff were literally running around the factory floors to deal with problems each time they arose. This situation consequently made smooth production an impossibility, at times resulting in overall stagnation.

Not too different were the earlier cases in Japan, where, in a significant number of instances, there was a partial introduction of sophisticated robots aimed at reducing the number of operational steps. Unfortunately, the performance of these robots was marred by minor but frequent stoppages. Before the production and maintenance departments could address the problems, the robots were discreetly removed from the plant floors. Even though recent models are indeed significantly better, both functionally and in terms of reliability, they are still likely to deteriorate under conditions of hard use. It is thus apparent that beyond a certain time span, the quality of maintenance and managements makes a marked difference in the rate of utilization of such technological innovations.

2.4 TOTAL PRODUCTIVE MAINTENANCE GOALS AND THE PROCESS OF IMPLEMENTING PRODUCTIVE MAINTENANCE DEPARTMENT GOALS

2.4.1 Development of PM Goals Closely Linked to Annual Plant Cost Reduction Goals

In the context of developing and utilizing total productive maintenance, we have described the ways in which the value of TPM can be understood by managers and plant supervisors. Its actual implementation in each enterprise should be adjusted in terms of the unique characteristics of the firm, such as the scale of its business, the size of each plant, product features, and differences in production modality. Of central importance, moreover, are the questions of whether the PM goal is consistence with production activities and whether the goal has been arrived at via a careful and thorough plan that can realistically be put into practice. Thus the crucial point in implementing the PM goal is the question of how it is linked to plant activities and how it can produce results in light of corporate and departmental plans for overall plant productivity improvement and annual plant cost reductions goals.

2.4.2 Preliminary Study

The usefulness of defining the goals and the ways of achieving them will differ significantly, depending upon how well they are conceived at the start. If the goals are only casually compiled from a variety of sources such as textbooks and seminar pamphlets, attempts at implementing them are bound to go astray.

The process of setting the goals and defining the specific implementation plans must be thoroughly analyzed in terms of actual circumstances, as well as the productivity improvement considerations that are relevant to the management and operation of each plant. Only then can each department follow the implementation procedures with a sense of commitment.

As part of the implementation plan, for example, standard operating procedure includes the rational organization or order of the plant and tools, tidiness, cleaning, personal cleanliness, and discipline. These are known in Japan as the "5S activities" (from the words: *seiri, seiton, seiso, seiketsu,* and *shitsuke*). Inspection of the facilities for cleanliness is also part of this plan. Ideally, the "5S activities" are to be practiced in every plant. It must be recognized, however, that the thoroughness with which these are actually implemented varies significantly among different plants and may often prove difficult to achieve in many of them. On the other hand, as long as plant supervisors are convinced of the necessity of these conditions, nothing less than their full implementation will be accepted. Moreover, when these activities are implemented by management, they are generally adopted by the workers. These "5S activities" will be discussed later as basic components in the promotion of PM (see section 3.7). Suffice it to say here that these elements can provide the diagnostic yardstick for measuring the extent to which PM implementation plans have been realized as part of a successful work environment.

It is often said that difficulties in implementation result from a lack of understanding at the top management level. It is important to note that productivity enhancement activities for lowering costs do not begin and end with PM alone. Depending upon the business environment, there are many instances in which measures must be quickly devised to produce faster results.

However, if workers realize that production cannot take place without their plant's machinery and facilities, and they start to believe in the many future benefits of PM, it should be possible for them to implement PM in their immediate plant environment, departments, and divisions. The people in top management should then respond positively and begin to understand the new methods.

By their very nature, these types of goals are not generally issued from the top management level; instead, they are created by the implementors and subsequently "sold" to top management personnel. If the top management people cannot be persuaded, it may be a result of a lack of one's own understanding of and confidence in the goals of PM.

2.4.3 Methods for Setting Specific Goals

To establish specific goals it is necessary to define every kind of loss that can occur when using machines, molds, and fixtures for each of the management checklist items (see Figure 2.3) relative to the output

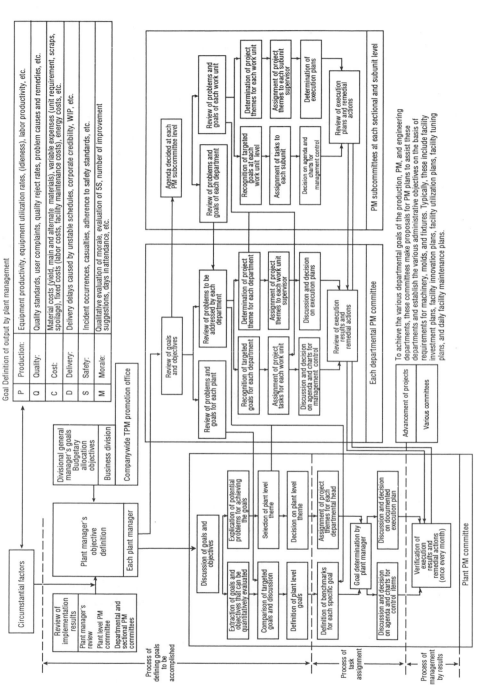

FIGURE 2.3 PM Goals and the Process of Implementing Productive Maintenance Department Goals

elements in plant management—P (production volume), Q (quality), C (costs), D (delivery dates), S (safety and environmental protection), and M (morale). As part of PM activities, the goals should incorporate the measures to be taken in this context.

The crucial point in determining the soundness of such plans is whether or not the PM activities are implemented only as repairs and restoration of machines and equipment, when they should be directly linked to plant productivity improvement in terms of quality, costs, and production management. As shown in Figure 2.3, what is important are the discussions and the ensuing consensus to be determined by the people involved in the process of defining the target objectives and assigning the tasks to attain these goals. In terms of PM, the question as to whether or not their followup analyses and remedial measures will be implemented in a systematic way depends upon making specific plans for carrying out PM for each department. The specific plans should focus on machines, molds, fixtures, gauges, and instruments and should include facility and machinery investment plans, facility and machinery improvement plans, plans for increasing the rate of equipment utilization for output, quality maintenance plans for the quality assurance programs, facility and machinery maintenance plans, and daily maintenance plans.

As a specific implementation example, Figure 2.4 shows the development of corporate goals and the basic TPM goals of a company that has accomplished significant results. Figure 2.5 shows the goals of management productivity on the basis of TPM.

2.5 THE ORGANIZATION'S ROLE IN ADVANCING THE USE OF TPM

2.5.1 TPM Advancement Forum and Specialized Committees

When organizing the companywide advancement of a TPM program, it is desirable to set up a general TPM promotion committee led by top management. If a TPM program is to be advanced on a plant-by-plant basis, then for each facility a plantwide TPM promotion committee should be set up as the nucleus, led by the plant's general manager and supported by an infrastructure consisting of TPM promotion groups for each section and subunit. The content of promotional activities at these levels must be specific; the most important aspect must be the organization of promotional units on the shop floor itself. This is where management skills and the leadership capabilities

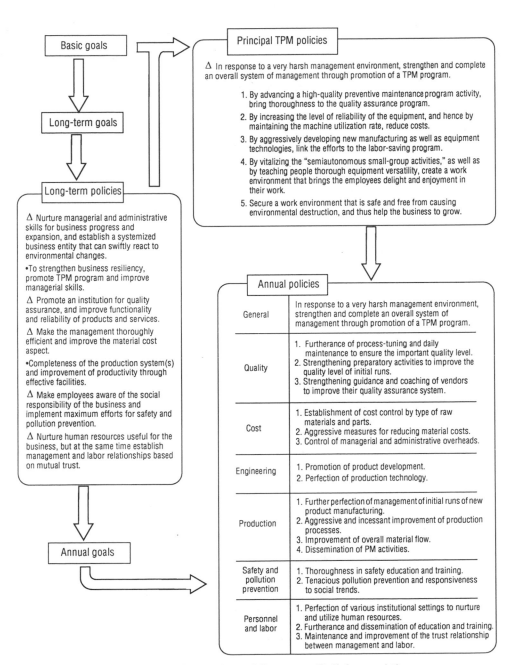

Basic goals

Long-term goals

Long-term policies

△ Nurture managerial and administrative skills for business progress and expansion, and establish a systemized business entity that can swiftly react to environmental changes.

•To strengthen business resiliency, promote TPM program and improve managerial skills.

△ Promote an institution for quality assurance, and improve functionality and reliability of products and services.

△ Make the management thoroughly efficient and improve the material cost aspect.

•Completeness of the production system(s) and improvement of productivity through effective facilities.

△ Make employees aware of the social responsibility of the business and implement maximum efforts for safety and pollution prevention.

△ Nurture human resources useful for the business, but at the same time establish management and labor relationships based on mutual trust.

Annual goals

Principal TPM policies

△ In response to a very harsh management environment, strengthen and complete an overall system of management through promotion of a TPM program.

1. By advancing a high-quality preventive maintenance program activity, bring thoroughness to the quality assurance program.
2. By increasing the level of reliability of the equipment, and hence by maintaining the machine utilization rate, reduce costs.
3. By aggressively developing new manufacturing as well as equipment technologies, link the efforts to the labor-saving program.
4. By vitalizing the "semiautonomous small-group activities," as well as by teaching people thorough equipment versatility, create a work environment that brings the employees delight and enjoyment in their work.
5. Secure a work environment that is safe and free from causing environmental destruction, and thus help the business to grow.

Annual policies

General	In response to a very harsh management environment, strengthen and complete an overall system of management through promotion of a TPM program.
Quality	1. Furtherance of process-tuning and daily maintenance to ensure the important quality level. 2. Strengthening preparatory activities to improve the quality level of initial runs. 3. Strengthening guidance and coaching of vendors to improve their quality assurance system.
Cost	1. Establishment of cost control by type of raw materials and parts. 2. Aggressive measures for reducing material costs. 3. Control of managerial and administrative overheads.
Engineering	1. Promotion of product development. 2. Perfection of production technology.
Production	1. Further perfection of management of initial runs of new product manufacturing. 2. Aggressive and incessant improvement of production processes. 3. Improvement of overall material flow. 4. Dissemination of PM activities.
Safety and pollution prevention	1. Thoroughness in safety education and training. 2. Tenacious pollution prevention and responsiveness to social trends.
Personnel and labor	1. Perfection of various institutional settings to nurture and utilize human resources. 2. Furtherance and dissemination of education and training. 3. Maintenance and improvement of the trust relationship between management and labor.

FIGURE 2.4 Expansion of Company Policies and the Principal TPM Goals

Pillar	Goal
1. 5S's and self-initiated maintenance	Organization of basic conditions and creation of a disciplined work environment
2. Human resources development (skills training)	·Production of workers well versed in manufacturing facility know-how and versatile in different skill requirements
3. Specialized maintenance (planning and management of maintenance)	Planned maintenance and improvements in maintenance technologies
4. Quality maintenance (building quality by means of facilities)	Elimination of chronic occurrences of defects and creation of QA lines (with 100% quality assurance)
5. Improvements in production efficiency and individual improvements	Giving visibility to losses, evaluation of efficiency and heightening the level of technological improvements
6. Equipment technologies (MP design and LCC)	Production innovations and vertical take-offs

FIGURE 2.5 Pillars of TPM Activities

of section chiefs and supervisors make the differences. Just as in the case of QC (quality control) group activities for quality assurance, the best method employs a mode of full member participation in which the PM groups themselves promote the program. The organization of these groups can be accomplished with relative ease. The contents of their activities, however, will vary broadly, ranging in practice from superb to meaningless, regardless of whether they are at a section or a group level.

The level of activities that the small work units can accomplish depends greatly upon the nature of the demands of the production process itself, on the worker–management relationship, and on the work environment. There are times, therefore, when it may be difficult to evaluate the group's activities. Nonetheless, the fact that the level of group activities within one work unit may be strikingly higher than that of others is generally a consequence of differing management skills of the supervisors of these different work units. Managers should possess the skills needed to motivate the activities of small work groups from within.

Regardless of the skill levels of specific managers, it is essential that top management communicate a sense of strong commitment to the PM program down through the lower echelons; this should be done via vertically organized promotional mechanisms such as the TPM promotion committee and departmental or sectional promotion meetings. It is also desirable to obtain the opinions of management staff members about the actual operational aspects and to take appropriate actions as needed. Topics for discussion include the following:

1. The degree of all-inclusiveness and specific goals of the PM plan for the plant(s).
2. PM implementation plans for the operational, maintenance, and staff function units.
3. Consistency and coordination of PM and evaluation of the PM group activities.
4. Promotion of awareness of PM and evaluation of the PM group activities.
5. Topics for PM education.
6. Assessment of monthly and quarterly overall maintenance performance and carryover plans for the next monthly and quarterly time frames.

Of special importance is the selection of a theme that will be emphasized in the project or projects aimed at a progressive and integrated implementation of activities involving maintenance, production, equipment planning, and unit design. To this end, it is desirable to set up some organizational units with horizontal mobility and flexibility, such as specialized TPM committees.

2.5.2 A System for Diagnosing and Evaluating the TPM Program and Its Operation

As part of an organized "closed loop," a system should be established for diagnosing and evaluating PM efforts. This allows for verification of the implementation of specific PM goals and enhances feedback as the goals are attained. The results in the following list can be achieved only through interorganizational diagnoses and evaluations from a companywide perspective.

1. The people in top management can have a more accurate understanding of the entire company and can therefore provide appropriate advice and directions.
2. The ideas and attitudes of top management can filter through to the rank and file, thereby raising employee morale.
3. The weak points of the companywide PM program will become more apparent in each plant, equipment, and engineering department. Increased awareness will motivate each department, the PM promotion committees, specialized implementation committees, and people in the maintenance coordination meetings to devise remedial measures.
4. Through the participation of each plant manager and interorganizational evaluators from various engineering and equipment units, the strengths and weaknesses of other plants and departments can be understood by all of these personnel from different parts of the company. This kind of crossover exposure can help limit deviation from PM standards within each of their organizational units.
5. Cooperation will be strengthened among plant managers and the engineering, equipment, and quality assurance departments, thus stimulating engineering remedies to problems with machinery and other technical equipment.

See Table 2.2 for a summary of the PM diagnosis, evaluation, and implementation system.

It is important to stress that all evaluators must learn the system and technologies of PM operations and maintenance, in order to raise their skill levels. Otherwise, efforts will soon be stymied and diagnoses and evaluations will become mere formalistic repetitions. Eval-

TABLE 2.2 Actual Implementation of PM Diagnostics

Name of diagnosis		Overall PM diagnosis (2-3 times per year)		Total company diagnosis (2-3 times per year)		Interdepartmental diagnosis (4 times per year for each department)
Goals of diagnosis		1. Level of progress in comparison with plan 2. Understanding of common and departmental weaknesses and advice (Period of transition from departmental PM method to TPM method)		1. Understanding the level of PM activities and advice on problem areas 2. Followup of items highlighted in previous diagnosis. (Perform PM diagnosis as part of the annual total company diagnosis; prior to start of a fiscal year)		1. Understand reality of implementing each individual item (projects) concerning PM activities and implement improvement measures 2. By mutually diagnosing each other's department, improve personnel skills and absorb other departments' strengths.
Object of diagnosis		Plant's departments		Entire division or plant		PM promotion departments
Diagnostic team members	Head diagnostician	President, PM promotional committee head, PM promotional committee assistant head	Head diagnostician	President, PM promotional committee head, PM promotional committee assistant head	Head diagnostician	Head of the PM committee office
	Diagnosticians	All departmental heads, plant manager(s), departmental PM promotional subcommittee heads	Diagnosticians	Departmental, divisional, and plant managers to take charge of quality, PM, and 5S aspects of diagnostics	Diagnosticians	Heads of departmental PM promotional committees; equipment supervisors
	Others	External consultants providing consulting services	Others	External consultants providing consulting services	Others	External consultants providing consulting services
Diagnostic methods	1. Schedule	3 hours per plant, 4 plants	1. Schedule	Functional department: 1.5 hours Plant: 2.5 hours	1. Schedule	1 department: 1.5 – 2 hours
	2. Steps	•Explanations of implementation realities of the plant •Individualized diagnosis; announcement of findings •Comments by each diagnostician; overall remarks by head diagnostician	2. Steps	•Explanations of self-evaluations of the PM implementation plan(s) •On-site diagnosis for plant only •Remarks by each diagnostician; overall remarks by head diagnostician	2. Steps	•Explanations by PM implementation subcommittee heads. •Individualized diagnosis by each diagnostician •Remarks by diagnosticians, reviews, and summary

TABLE 2.3 TPM Total Diagnostic Evaluation Sheet

(STEP-1)

	Diagnostic category	Diagnostic points	Appraisal (O, △, X)	Remarks
1	Method of TPM implementation	1) Whether or not top management's goals and requirements are clearly reflected		
2	TPM implementation system			
3	TPM enlightenment status			
4	Equipment assignment system			
5	Status of 5S — Neatness-Tidiness			
6	Status of 5S — Cleanliness-Cleaning			
7	Lubrication management			
8	Overall evaluation			
	General remarks			

(STEP-2)

	Diagnostic items	Diagnostic points	Evaluation (Mark with a circle O at an appropriate value in the scale for each item.)	Remarks
1	Goals of TPM implementation (8 points)	1) Whether or not current problems and their remedial actions are specific.	Specific 4 — 3 — 2 — 1 Abstract	
		2) Whether or not target date and specific target objectives for remedies are clear.	Clear 4 — 3 — 2 — 1 Obscure	
2	System for TPM implementation (24 points)	1) Whether or not frequency and contents of group PM meetings are substantive (examine proceedings).	More than once a month PM meetings / Substantive contents 4 — 3 — Insubstantial contents 2 — 1	

(STEP-3)

	Diagnostic item	Diagnostic points	Evaluation (mark with a circle O at an appropriate value in the scale for each item.)	Remarks
1	TPM implementation goals (8 points)	1) Whether or not current problems and their remedial actions are specific.	Specific 2.5 — 2.0 — 1.5 — 1.0 Abstract	
		2) Whether or not target date and specific target objectives for remedies are clear.	Clear 2.5 — 2.0 — 1.5 — 1.0 Obscure	
2	Training people (7.5 points)	1) Whether or not frequency and contents of group PM meetings are substantive (examine proceedings).	Concrete 2.5 — 2.0 — 1.5 — 1.0 No Goals	
3	Reduction in equipment breakdowns (21 points)			
4	Reduction in setup time and stoppages (21 points)			
5	5S's of equipment (16 points)			
6	Lubrication accidents (10 points)			

uators should receive advice from capable consultants, attend seminars and other educational activities, and continue to sharpen their skills.

The manager and supervisor in charge of the unit being evaluated should attend all diagnostic and evaluation sessions. Evaluation topics are shown in Table 2.3. All personnel in the unit being evaluated, not only the diagnostic and evaluation team, should meet together formally and share the evaluation results. The skill of each evaluator will thus be improved, and appropriate advice will be communicated to the unit. The group may also be inspired to improve upon its own performance by perceiving the strengths and weaknesses of other work units. To obtain even better results from this process, the format used in the summary of the official announcement should also be standardized.

3

Analysis and Diagnosis of Management Standards for PM and Methods for Improvement

3.1 ANALYSIS OF CURRENT REALITIES OF PM ACTIVITIES AND MANAGEMENT STANDARDS

In order to perceive and evaluate accurately the PM activities in a plant, it is necessary to investigate the realities of the ongoing processes, without being confused by organizational setups or management systems. At this stage it is important to be able to accept the realities as they occur, without bias, and without being caught up in established notions about what they are. In other words, the situation should be analyzed from an objective point of view, rather than having conclusions drawn based on convenience.

There are plants that seem to have a fine management system in place, but in reality a situation may exist whereby personnel are in a state of continuous pressure, spending their time chasing after inconsequential "busy work." On the other hand, there are plants whose management systems are not initially impressive and which offer rather naive explanations of how their systems operate, yet these same plants show dedicated adherence to the basics of PM with a potential for a tremendous leap in productivity once the PM principles are understood.

Thus, to find a clear answer to the question "What is the substance of the problem?" it is necessary to define the production-related problems resulting from the condition of the equipment. The equipment-related problems, moreover, must be analyzed in light of the previously discussed output elements in plant management, i.e., P, Q, C, D, S, and M.

3.2 ANALYSIS OF THE RELATIONSHIP BETWEEN PRODUCTION (P: PRODUCTION VOLUME) AND EQUIPMENT

—BUILDING PRODUCTION LINES WITH HIGH PRODUCTIVITY AND
ZERO BREAKDOWN—

How effective are the investments made for expansion and improvement of equipment? Are the effects of these investments realized to the degree anticipated during budget planning, within the context of intended equipment automation, mechanization, and labor reduction? There have been many cases in which investments that were intended to reduce labor needs failed to meet the anticipated goals because of problems with the equipment, producing only unclear results with regard to efforts to reduce operational steps. The realities of equipment utilization must therefore be investigated. Different industries have different rates of equipment utilization. The typical equipment utilization rate is found in an environment of individualized piece production, where the manufactured product consists of a mixture of many kinds of products in small quantities. In contrast, a mass-production parts assembly factory typically has a different rate. An equipment-dominated processing industry typically has yet another process utilization rate. The criteria to be applied in these varying environments will obviously be different as well. Nevertheless, the equipment utilization rate during the time period for which the production schedule has been planned should register a rate of close to 100 percent.

The question should also be investigated as to whether the equipment and the production lines are running at a sufficient level of capacity relative to their standard capabilities as called for in their specifications. Are they running unnoticed at slower rates than they should be? Are their reject rates too high, even though their operational rates seem to be adequate? The index derived from the losses from these problems is called the rate of equipment effectiveness (or rate of equipment utilization for output). (See Figure 3.1.)

PM is a management technique that challenges the accepted notions of equipment efficiency. When the real condition of a plant is to be diagnosed, we ask that the major equipment be defined, and we gather data about their utilization rate, the size of the inventory of finished goods, and the work-in-process inventory. If these questions

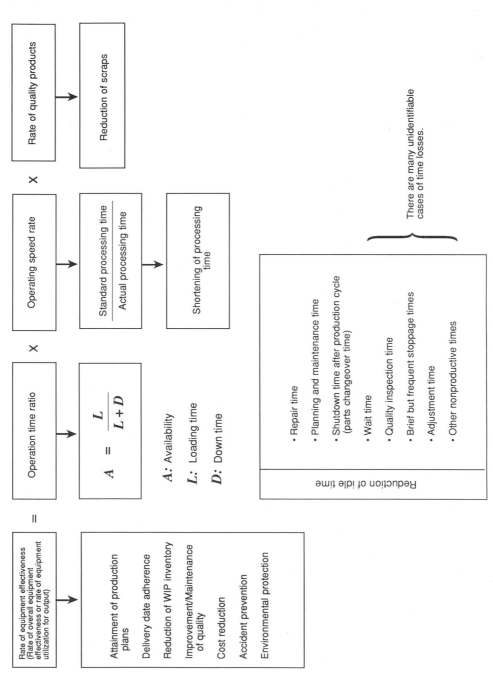

FIGURE 3.1 Rate of Equipment Effectiveness (Rate of Equipment Utilization for Output)

are addressed directly to the personnel in charge of the equipment, answers are often noticeably absent. These personnel tend to regard the equipment utilization rate of each operation as a matter of concern for production management and hence outside the scope of their management agenda. Even among some echelons of manufacturing management, moreover, we sometimes observe people making presentations using one of their management monthly reports. It is clear that they are collecting data but hardly managing.

If we are to appropriately question and challenge the data with regard to quality management, equipment utilization, work in process, or costs, then we must be thoroughly familiar with their target statistics. By memorizing these statistics, we can immediately respond when someone asks about them. If the front-line supervisors of the manufacturing processes or the maintenance staff do not know the utilization rates, however, one might conclude that PM is not actually present.

Equipment will undergo different utilization curves depending upon the load profiles of the shop floor production schedules. That is to say, the changeovers of the molds and rollers that are needed for each of the various product types would increase in frequency, and concomitantly, the total time of stoppages, in the aggregate, would also increase. However, the molds and rollers may be of concern to the manufacturing department, since they are important parts of its production equipment. But in reality, the industrial engineering department—as the equipment-designing unit of the plant—should not and cannot afford to avoid its responsibility to strive toward productivity improvement. This department must treat such fixtures and molds as important elements and the key to furthering the goal of integrated productivity design. At the same time, the equipment engineering department or the maintenance department must make it their own goal to improve upon the rate of equipment utilization for output. It is not just a matter of maintenance by the manufacturing department alone.

From the perspective of historical change in equipment investments in industries and in the rate of utilization for output, Japanese industries have, by and large, adopted strategies based on a management philosophy that is a type of market share expansionism. This has often led to a phenomenon in which the growth in production capacity outpaced that of market demands. It has been one of the factors that has discouraged serious efforts at investigating the utilization rates of existing equipment. This tendency might have promoted casual equipment improvements and expansions. So long as there is a promise of ongoing expansion of market shares, it may be a natural course of action for management to continue expanding capacities by adding new equipment and adopting revolutionary and new technologies. But implicit in this approach are many unknowns with regard to prediction of the future usefulness of equipment. Moreover, the amortization burdens and the interest payments would not make this approach a financially viable one on a short-term basis. Depending on the characteristics of the products involved, this point may become quite pertinent, i.e., in the case of products that are manufactured and sold as means of production.

By and large, profitable plants are relatively old ones that have few interest payments or amortization burdens. When we study each piece of equipment for each of their operations, a significant number of partial expansions and innovations to the equipment becomes noticeable. Moreover, efforts in reducing labor and increasing capacity have been accomplished through low-cost ways and means. From the standpoint of investment for renovation, it is desirable to have a positive attitude toward the introduction of new equipment together with technological innovations. But we must not forget to first investigate ways to utilize existing equipment to its maximum potential, e.g., by exploring the possibilities of increasing capacities and useful life span by improving existing equipment.

Equipment does not break down simply because it is old. Of course, it is difficult to expect an equal level of function and precision in older as compared to newer equipment—yet we can certainly maintain the functioning and precision levels of the original design specifications in older equipment. One can find numerous case studies of this in developing countries that cannot yet utilize the newest equipment. There are many instances in which used equipment has been well renovated, equipped with spare parts, and exported to these countries. The higher level of labor-intensive activity required to maintain and operate this renovated equipment and its lower rate of production are not the problems in developing countries. Rather, so long as these machines can maintain expected functioning and precision levels, many of them can serve useful purposes quite well. This phenomenon simply points out the fact that equipment does not break down because it is outdated, but because people think that outdated equipment will break down of its own accord, and thus they do not maintain it as assiduously as they would maintain new equipment.

A perception of future trends in the Japanese economy strongly indicates that there is an acute need to curtail capital investments other than those that are truly necessary, to maximize the use of existing equipment, and to reduce basic manufacturing costs. From its very inception, PM has been a management technique that takes on the challenge of improving equipment efficiency. This being the case, as long as there exists excessive capacity in equipment, management would tend to focus on such areas as the prevention of quality-reject items resulting from equipment-related problems, the improvement of energy conservation measures related to equipment, and the reduction of equipment maintenance costs. There has recently been a rapid transition in the means of production from a system based on people to one based on equipment. Given this type of manufacturing environment, if we strive to suppress equipment investments and still try to meet increased production demands, then production planning means only the planning of machine utilization schedules. This is the most economical stoppage plan of equipment itself for minimizing quality rejects that may result from unexpected breakdowns and equipment malfunctions. In other words, the key question here is how the reliability of equipment can be integrally planned together with the production plan, so as to increase the effectiveness of equipment utilization on an annual basis.

As a rule, equipment does not break down all that easily. But in production processes in which much machinery is involved, unless the reliability of each component is managed as part of the overall reliability of the entire group of machines involved in the manufacturing operation, the mean time between failures of those machines will be completely unsynchronized and unpredictable. Thus, breakdowns may be occurring at any time in one or more operations, thereby causing the utilization rate of the entire facility to decrease dramatically. For this reason, the optimum checkup cycle for each individual machine should be studied and defined. At the same time, the economic viability of batched maintenance plans for each of the groups of machines with different mean times between failures should be studied for all the components involved in one whole cycle of the production process.

Assessment of the value of a synchronized maintenance plan should be made in light of a trade-off between (1) the aggregate cost of extraneous maintenance costs incurred from the maintenance of those parts whose mean times between failures would fall between the regular maintenance of other machinery and (2) the aggregate of the maintenance-derived savings of the costs that would be incurred from unexpected stoppages caused by unscheduled downtimes. It should be noted that the more diversified the total equipment becomes and the greater the number of machines installed, the more difficult the determination of an economical stoppage plan will become.

Given the variations in usefulness that depend upon plant size, the determination of such an economical stoppage plan must be computer controlled in the future, especially in view of the plan's interrelationship with marketing plans. Moreover, to make such plans there is a need to investigate inspection technologies further for better predicting the life span of crucial areas of the equipment. With regard to inspection technologies, in cases where it is technically very difficult to estimate life spans, we should be aware of the need for extrapolating predictable life spans on the basis of reliable data derived from the past statistics of breakdowns. In order to collect and use this type of data accurately and reliably, the united cooperation of the production, maintenance, repair, and design engineering departments is a highly desirable goal.

In the area of equipment-related technology and techniques, many important problems may engage a considerable number of professionals who may often feel that they lack the necessary skills to cope with them. These problems pose no dearth of challenges. Even for a machine that at present may be running smoothly, the problem of prolonging the life span of its important components is obviously an important challenge. Moreover, the technology of equipment is comprised of cumulative elements of technological knowledge, and its improvement is hardly simple or straightforward. Individual initiative, in particular, makes a major difference. Thus, instead of simply trying to "make do" with the application of the existing technology as supplied by the equipment vendors or through application of the standard reference materials, it is desirable to cultivate homegrown, self-initiated skills and techniques, even those seeming to be of little significance.

When we examine the efficiency of equipment operations as a management index, the rate of failure occurrences and the rate of extended service are important factors to be considered. Moreover, the equipment idle time required for changeover setups is taken into consideration as the changeover times for fixtures, rollers, molds, nozzles, and so on, and these times are required to be compressed. They constitute a subject for investigation that is indispensable for machine parts maintenance. At the same time, the rate of adherence to planned maintenance should be reviewed in light of the rate of unexpected failures and that of extended service life.

In the following section, the improvement of efficiency of machine operations will be discussed, i.e., the means for promoting development of production lines with zero unexpected breakdowns and high throughput productivity level.

3.2.1 Rate of Equipment Effectiveness and Management of Time-Stamped Output per Time Period

In assembly type manufacturing, some plants coordinate setups and staging by means of diagrams. This assists in the improvement of operational steps and equipment, enabling completion of each operation within the allotted time. In practice, however, operational time is often significantly overrun as a result of short but frequent production stoppages and unaccounted-for "time outs." To address these problems, the goal of completing work within allotted operation time cycles can be promoted through improvement measures in each area of production. These measures include correct assessment of output efficiency based on the diagrammed plan for coordinating setups and staging, as well as explicit identification of any hidden or unexplained nonoperational times. In plants where output efficiency is not considered a major indicator of productivity, machine utilization is only about 50 to 70 percent, depending upon the frequency of changeovers and on each of the setup times. Unless this rate is raised to 85 percent or higher, however, output efficiency is low.

PM is comprised of activities that take on the challenge of improving the rate of equipment effectiveness. As can be surmised from the equations shown in Figure 3.1, some PM activities are aimed at increasing the capacity of equipment and production lines, and others are directed at reducing wasted operation times. If such waste is reduced, equipment capacities obviously will be increased. In terms of PM applications, these activities will enhance capacity in a more positive manner, overcoming the tendency to avoid dealing directly with the realities of wasted operation time and the extent and total impact of the problems. Figure 3.2 shows the points at which improvements can be made.

In Figure 3.1, the rate of equipment effectiveness is given as:

(Rate of Equipment Effectiveness) = (Operation Time Ratio)
$\times$ (Operating Speed Rate)
$\times$ (Rate of Quality Products)

For major equipment, this equation should be applied using the detailed breakdowns as shown in the rate of equipment effectiveness formula. Generally, however, the formula should be approximated to the rate of utilization for output; in many cases this abbreviated version is easier to use.

The next question to be addressed is productivity improvement. There are two aspects to improving productivity: the first is improve-

Goals		Improvement points
Improvement in rate of equipment effectiveness	Increase in equipment capacity (lines)	• Speed increase, operational simplification, etc. • Single-action process: increase in functional and precision levels • Increase in batch size; high volume processing, etc. • Use of excess capacity; line load balancing through "smart" layout • Extended operation time; increase in the number of machines • Unattended operation • Mechanization of manual work • Others
	Reduction in operational losses	• Reduction of breakdowns • Reduction of brief stoppages and idling time • Reduction in setup times • Reduction in scraps and reworks • Reduction in wait and queue times • Reduction in other losses

FIGURE 3.2

ment in the production efficiency of equipment and the second is improvement in labor. On the production floor, both aspects should be maximized. Whereas the former can be considered as improvement in rate of equipment effectiveness, the latter has been considered part of the activities for enhancing overall work efficiency, such as reduction in the number of operational steps. These activities call for different emphases, depending upon the specific requirements and local needs of the shop floors.

In Figure 3.3, a typical example is shown dealing with automated production lines. The methods for increasing productivity in automated production lines can be summarized as improvement in the rate of equipment effectiveness (rate of overall equipment effectiveness). On the other hand, manual production lines consist of the combined manufacturing activities of labor and machines or tools. In this kind of manufacturing line, the operational layout is not designed to make the equipment run at a 100 percent utilization rate. Rather, the layout is designed to keep people occupied by work 100 percent of the time, and the production flow lines are conceptualized along this kind of thinking. As a result, some machines become bottlenecks. There are cases in which improvement in the effective utilization of such "bottleneck" machines is important. At any rate, manual-labor-oriented

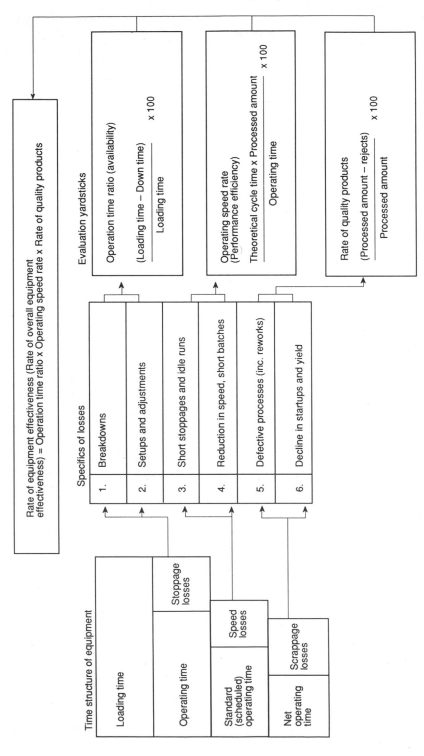

FIGURE 3.3 Rate of Equipment Effectiveness

Rate of equipment effectiveness (Rate of overall equipment effectiveness) = Operation time ratio x Operating speed rate x Rate of quality products

Evaluation yardsticks

Operation time ratio (availability)

$$\frac{(\text{Loading time} - \text{Down time})}{\text{Loading time}} \times 100$$

Operating speed rate (Performance efficiency)

$$\frac{\text{Theoretical cycle time} \times \text{Processed amount}}{\text{Operating time}} \times 100$$

Rate of quality products

$$\frac{(\text{Processed amount} - \text{rejects})}{\text{Processed amount}} \times 100$$

Specifics of losses

1. Breakdowns

2. Setups and adjustments

3. Short stoppages and idle runs

4. Reduction in speed, short batches

5. Defective processes (inc. reworks)

6. Decline in startups and yield

Time structure of equipment

Loading time

Operating time

Standard (scheduled) operating time

Net operating time

Stoppage losses

Speed losses

Scrappage losses

52

manufacturing lines should be considered as a type of production line combining people and machinery. For this type of production line, it is necessary to think in terms of improving the production capacity of such a combined production line as a whole; thus one must approach the problem from a viewpoint that takes into consideration more than the efficiency of the physical equipment alone.

Successful efforts to increase output depend on whether three key questions are thoroughly understood: (1) What is the maximum capacity of the production line or the equipment? (2) What is the target value? (3) What are the current capabilities? With regard to fluctuations in output of the main production line(s), it will be difficult to achieve significant results unless the situation is analyzed in detail and remedial measures are taken using a time-period-based output management table.

After a main production line has been selected for improvement, the following measures should be taken:

• Establish a series of time-period-based target output values.
• Keep precise records for each time period. Record each worker's name. To raise morale, indicate those workers who achieved the desired goals or established a record output.
• When problems arise, the worker should report to his or her superior(s) immediately so as to minimize the mean time for repairs.

Using an output table with the breakdown for each of the time periods, some typical problems include the slowness of getting work underway in the mornings (especially Monday mornings) or upon resuming work after lunch. In many cases, these problems may be remedied by the use of a diagram of staggered work-reporting schedules among a small number of people. This approach would also pinpoint idle times caused by delays in the delivery of parts from preceding operations. This in turn would facilitate taking adequate measures to solve these problems.

Fundamental to this approach is the laying of a foundation that would smoothly facilitate detailed work management at the individual worker's level. It is crucial to establish an environment in which an individual worker can comfortably initiate reporting or analysis of problems and suggest small and detailed improvements. This may be the most critical test, as it were, of the mettle of a manufacturing plant. For an example of a table with time periods for output management, see Figure 3.4.

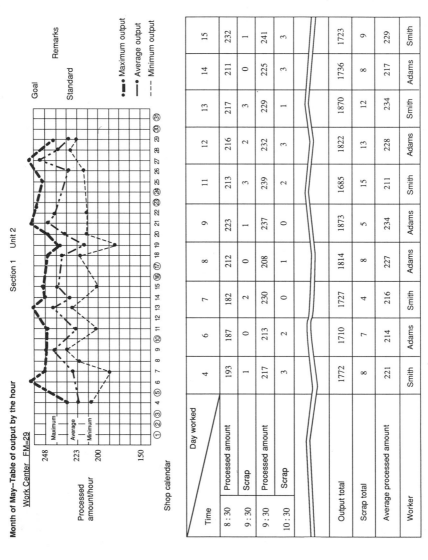

FIGURE 3.4 Table of Output by Line

3.2.2 The PM Approach and Improvement in Rate of Equipment Effectiveness

1. EQUIPMENT AND CHARACTERISTICS OF OPERATIONS

As a rule, there are few significant differences in the objectives or orientation of PM activities between process industries and machine-processing and assembly industries. But significant differences do exist with regard to the implementation of PM activities, for example, between (1) process equipment such as continuous electrolytic plating equipment, electric arc furnaces, or continuous painting equipment; and (2) fabrication and assembly equipment (dedicated or general purpose types) such as pressure-sealing pipemaking machines, pneumatically operated automatic assembly machines, welding machines (of torch, arc, or carbon-dioxide types), or single-task milling machines. The mode of operations and the way equipment is used are very different in each of these cases. They may be run for six months or for one week continuously, or for eight hours per day, or in sixteen-hour operation cycles. Because of these variations, a different maintenance method should be established for each of them and the most economical method should be adhered to. Depending upon the kind of equipment, required maintenance skill levels may also be different.

The initial goal following the introduction of TPM is the total elimination of equipment breakdowns. As noted earlier with regard to organization for TPM promotion, committees are created (specializing in MTBF analysis, improvement in the rate of equipment utilization for output, and management of lubrication), for the purpose of achieving systematic equipment maintenance. These special committees should then play pivotal roles in selecting a pilot production line, educating and training in MTBF analysis, coping with problems of breakdowns with long downtimes, and strengthening the capability of the maintenance systems.

The maintenance department should strive to increase the effectiveness of activities based on a maintenance calendar. The manufacturing department should nurture the personnel who can assume leadership in PM activities; while using these people as a nucleus, the department should also encourage self-initiated productivity maintenance activities.

2. TYPES OF BREAKDOWNS AND RESULTANT LOSSES (FIGURE 3.5)

Breakdowns and malfunctions can generally be divided into either an abrupt breakdown or a deterioration breakdown type. Abrupt break-

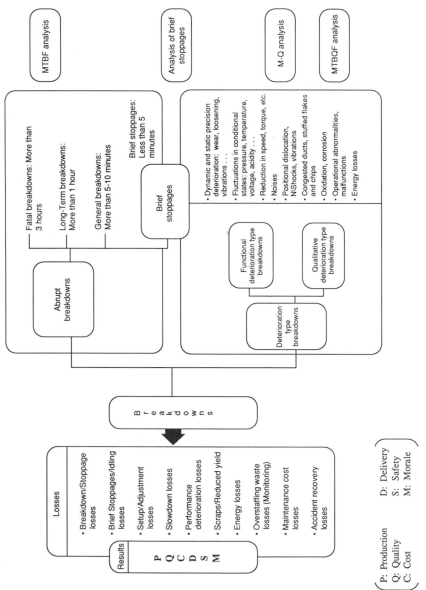

FIGURE 3.5 Types of Breakdowns and Losses

downs can be divided into different subcategories, depending upon the length of stoppage times, such as (1) fatal breakdowns, (2) long-term breakdowns, (3) general breakdowns, and (4) frequent but minor breakdowns. Breakdowns caused by equipment deterioration or aging can be subdivided into functional deterioration and output quality type breakdowns. In the case of functional deterioration, the equipment in question functions properly, but its ejection quantity or (more generally) its output quantity gradually decreases, impacting on the amount of related work-in-process inventory or on the lot sizing of the materials involved.

In machine processing and assembly industry, the quality deterioration breakdown involves deterioration of a portion of the whole production facility, to a degree that quality reject output occurs before anyone notices it. The losses caused by these breakdowns may include breakdown stoppage losses, idling losses caused by minor but frequent breakdowns, losses caused by schedule adjustments, production speed slowdown, reject WIP inventory, maintenance costs and natural calamities, and losses resulting from the number of operational steps.

Ironically, it is often difficult to convince people that breakdowns cause such a variety of losses and that the elimination of breakdowns means improvement in productivity, quality, and cost effectiveness. Why is it so difficult? The answer can be found in examining the records of operations stoppages (when work could not be performed), which should be kept on all shop floors. Figure 3.6 shows a Pareto diagram of various times that are included as part of a machine's operational steps, other than those used for actual machine runs. In this diagram, the size of breakdowns is small and relatively obscure; this

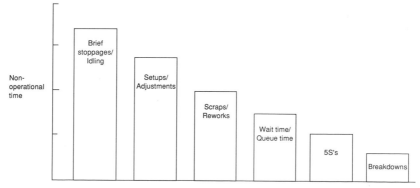

FIGURE 3.6 Pareto Diagram of Nonoperational Time

denotes short stoppages. Such stoppages are usually known as dry-run times or unclear usage times, and, in general practice, they are considered unrelated to equipment malfunctions.

This Pareto diagram could be interpreted as showing the need to tackle the reduction of setup times and a reduction in scraps. Let us, however, consider the causal relationship between setup time and scraps. For example, the factors contributing to generation of scraps would point to cases where something went wrong with any of the equipment, personnel, materials, or methods. In particular, there are many cases of product quality reject situations that are caused by equipment malfunctions. The remedy for these situations may be repair of the malfunctioning equipment. However, the net result is that scraps are generated, thereby lowering the level of productivity. Thinking along this line, we may perceive problems other than scraps that also have much to do with equipment breakdowns or malfunctions. Even so, from the day-to-day management data, the importance of properly functioning equipment would not necessarily be perceived. This is precisely why it is necessary to recognize and remind others about the importance of equipment and of ways of collecting data to facilitate such awareness. Thus, activities aiming at thoroughly eradicating losses could be described as breakdown prevention activities with total employee participation.

As discussed before, the subject of productivity improvement should be approached from a perspective based on understanding of the relationship between equipment breakdowns and malfunctions and productivity loss. Table 3.1 shows the interrelationships among various factors that hinder productivity for automated fabrication lines. Subjects addressed include (1) breakdowns, (2) minor but frequent stoppages, (3) scraps, (4) setups and changeovers, (5) increases in speed, (6) unattended operation, and (7) extensions of lathe bit life. Any one of these must be approached within the PM framework.

To prevent breakdowns and malfunctions of equipment, improvements and modifications must be recognized as having an important relationship to PM. With regard to assembly lines, these phenomena are related to minor but frequent stoppages, component shortages, scraps, and reworks, which all result in losses in labor productivity. For the assembly line environment, it is essential that measures devised to counter minor stoppages, procedures for adjustments, and techniques focusing on increases in human skill levels and error prevention should also incorporate the PM approach.

TABLE 3.1 Topics for Production Efficiency

Automated process line	Product group	RY1010Y

○: Already in place
△: Not yet in place
✕: Not needed

No.	Line identification	Production volume/day	Load hours/day	Cycle time/sec	Operational contraction	Speed increase	Machine used	Time in minutes	Stock	Time in minutes	Drop in main shaft's RPM	Ceramic blades	Size errors and stopper change	Mislocation of mounting holes: installation of touch-sensitive switch	Antiswing measures: improvement in bushings	Changeovers	Break in chuck/cylinder	Break in drill bits	Measures to be taken for razor blades	Break in holder	Oil leakage	Sawdust cover	Catching duct shoot	Redo of setup	Overruns
1	MA1101	20000	23.5	4.23	○	△	△	14.0	△	8.5	✕	△	○	○	△	△	○	△	○	✕	△	△	○	△	△
2	MA1102	17000	12.5	2.65	△	△	✕	✕ 20	✕	✕ 16	○	△	△	✕	△	△	△	○	✕	△	△	○	○	✕	○
3	MA1103	11000	14.0	4.58	○	✕	○	13.5	✕	✕	✕	△	✕	○	○	✕	○	○	△	✕	○	✕	✕	△	○
4	MA1104	22000	20.8	3.40	○	△	○	6.5	△	5.1	✕	✕	○	△	△	✕	△	△	○	○	△	△	○	✕	○
5	MA1105	7800	10.0	4.40	△	✕	✕	✕	○	3.5	✕	✕	✕	△	✕	○	✕	✕	△	✕	△	✕	✕	△	✕

Column groupings:
- Work load: Production volume/day, Load hours/day
- Speed: Cycle time/sec, Operational contraction, Speed increase
- Unattended operation — Material: Machine used, Time in minutes
- Unattended operation — Finished goods: Stock, Time in minutes
- Extension of bit life: Drop in main shaft's RPM, Ceramic blades
- Measures against scraps: Size errors and stopper change, Mislocation of mounting holes: installation of touch-sensitive switch, Antiswing measures: improvement in bushings
- Changeovers
- Measures against breakdowns: Break in chuck/cylinder, Break in drill bits, Measures to be taken for razor blades, Break in holder
- Measures toward 5S's: Oil leakage, Sawdust cover
- Brief stoppages: Catching duct shoot, Redo of setup, Overruns

59

3.2.3 Prevention of Breakdowns Through Total Employee Participation

The extent of a breakdown generally depends upon the production characteristics and size of the equipment in question. In mechanical-assembly industries, most sudden breakdowns can be repaired within an hour or less. If efforts are made to analyze the specifics of those breakdowns and implement a planned productive maintenance program, the amount of time needed to cope with sudden breakdowns can be cut in half or even more.

For example, if lunch breaks are staggered among the maintenance crew members, breakdowns will decrease. Moreover, if a program of staggered "flex-time" work hours is implemented to carry out regular maintenance activities for three hours on a daily basis over a period of two to three months, the number of breakdowns should again noticeably decrease. Since maintenance activities can be carried out only when machines are stopped, the maintenance crew must function during stoppage times. The shop calendar of the maintenance crew, therefore, will be different from that of the production operation crew. Unless this flexibility is implemented, the maintenance department will be required to work overtime day after day, until the entire department crew is too tired to continue a satisfactory program, despite the fact that production operations must and would continue. In many cases, breakdowns would then occur. Traditional work schedules based on set patterns, therefore, should be liberally modified so as to provide the versatility needed to meet operational requirements.

A total employee participation program of daily maintenance—based on the idea of minor preventive maintenance—is made possible by staggered lunch break schedules among the maintenance crew members, as well as by self-initiated maintenance undertaken by the operational department units. This makes it possible to reduce breakdown occurrences to one-half—or even one-fifth or one-tenth—of previous levels. We maintain that there is a substantial difference between a reduction of total plant equipment breakdowns by ten or twenty percent and a reduction to one-fifth or one-tenth of the original occurrences. To bring the breakdown occurrences to one-tenth the current rate means that the qualitative characteristics of the equipment have been altered. This result is obtainable only if the behavior of everyone involved in handling the equipment is significantly changed.

The method of maintenance adopted must be suited to the characteristics of each individual piece of equipment and to its production modality. Most importantly, achieving any satisfactory level of productive maintenance is possible only if measures are taken to ensure that when greater production demands are placed on the equipment, efforts to prevent production stoppages are more effective. Once a stoppage does occur, whether because of production changeovers or sudden breakdowns, the stoppage time should be put to good use. Thus, there should be a maintenance plan prepared in advance for those items that are due to be inspected and maintained in their regular maintenance cycles. Whenever a stoppage occurs in any operation involving these items, appropriate maintenance actions can then be taken at the same time for all of the items involved. Without this overall practice, reduction of the breakdown rate will be impossible to achieve. Because of demands on production schedules, requests for machine stoppages for maintenance reasons might be overridden, thereby limiting the performance of maintenance to those times when stoppages occur because of breakdowns. Thus, it is necessary that production and maintenance teams keep in close touch with each other and collaborate on a detailed and planned maintenance program so that both maintenance teams will (1) understand the realities of the equipment utilization plan, (2) have time estimates of planned maintenance operations, (3) make replacement parts available, and (4) prepare different maintenance plans appropriate for stoppages of varying length, whether thirty minutes or two hours. Unless this kind of coordination is accomplished, maintenance activities cannot closely support the goals of the production team.

1. SELECTION OF EQUIPMENT FOR CRITICAL PM MANAGEMENT AND PRIORITIZATION METHOD

In order to maximize the efficiency and effectiveness of PM activities, critical equipment (i.e., production lines) must be understood in terms of (1) current production environment, (2) limited human resources, and (3) costs. In some industries, critical equipment or production lines tend to change drastically, depending upon production and quality demands. In those industries where significant changes occur, a ranking table should be devised to allow a simplified equipment ranking to be updated in a review cycle of about once every six months.

"Critical equipment" can be defined as equipment or a production line that requires critical management with regard to production and quality, in order to increase the effects of productive maintenance. From the viewpoint of production, this term refers to (1) equipment experiencing frequent breakdowns, (2) production equipment with no backups or reserve capacities, (3) equipment that would significantly reduce total throughput if it broke down, (4) equipment whose breakdown would affect the due dates of many items, (5) equipment that is close to the finishing phases of the production processes, (6) equipment that would affect production timing (i.e., highly loaded production lines), or (7) equipment whose breakdown would cause delays in overall production.

From a quality perspective, critical equipment is (1) that which greatly affects product quality, (2) that which causes quality level variation, or (3) that which would cause variation in quality level when breakdowns occur. Of course, it is also important to evaluate equipment on the basis of magnitude of maintenance cost, but here we are seeking to establish a set of evaluation criteria based on quality and breakdowns. The cost and safety implications of maintenance should be approached using different evaluation criteria.

For a table of equipment assigned priorities from the PM management viewpoint and focusing on fabrication-type equipment, see Figure 3.7.

2. ANALYSIS OF BREAKDOWNS AND REMEDIES

Analysis of a breakdown should start with investigations of the various factors surrounding the breakdown, enabling detailed analysis of each factor. Likewise, a comprehensive approach should be taken to find the remedy. The first elements involved in the breakdown to be analyzed include identification of the following questions: (1) Which equipment belonging to which manufacturing team(s) are sustaining the most frequent breakdowns and who is responsible for the team(s)? (2) Which process(es) or line(s) are registering the most frequent breakdowns? (3) What kind of equipment sustains the most frequent breakdowns? (4) Which functions or structures of the equipment sustain the most frequent breakdowns (for example, the positioning mechanisms, the removal and unloading mechanisms, the alignment mechanisms, the holding or grasping mechanisms, and the workpiece rotation mechanisms)? (5) Which specific areas of these mechanisms break down most frequently, or which occur in the areas of the main chassis or frame section of the machine—the power mechanisms, the drive train mechanisms or the fixtures? Moving a step further into the specific component level, which part or component breaks down most frequently? Is it a cylinder, a clamping fixture, a joint assembly, a tubing fixture, a bearing, a shaft, a belt, or a spring?

Aside from the different levels of repair details, repair and maintenance in every plant are carried out each time a sudden breakdown takes place. But the road toward "zero breakdown" lies in the idea of total employee involvement in equipment breakdowns, and requires

Table of assessment of high priority equipment (production lines) for TPM					Month investigated _____

Facility/Line			Operation ID	Facility Line ID			Machine no.

Type		No.	Item	Assessment				Assessment criteria
Production	Production aspects	1	Frequency of occurrences of abrupt breakdowns		4	2	1	④ More than 15 occurrences/month ② From 14 occurrences/month to 6 occurrences/month ① Less than 5 occurrences/month
		2	Frequency of long breakdowns		4	2	1	④ More than 4 occurrences/month ② From 3 occurrences/month to 2 occurrences/month ① Less than 1 occurrence/month
		3	Mean time to repair (MTTR)		4	2	1	④ More than 2 hours/occurrence ② From 2 hours to 1 hour/occurrence ① Less than 1 hour/occurrence
	Delivery aspects	4	Impact of breakdown on the preceding and following operations (for fatal breakdowns)	5	4	2	1	⑤④ Breakdown affects production of 2 or more lines ② Breakdown causes 1 line to come to a halt ① Breakdown causes reduction in capacity but allows continuation of operations
		5	Impact of breakdown on customers	5	4	2	1	⑤④ Breakdowns causing line stoppages in customer production equipment ② Breakdowns causing depreciation of safety stocks and subsequent line stoppages ① Breakdowns causing difficulties or problems in subsequent lines
Quality		6	Occurrences of quality rejects		4	2	1	④ More than 0.1% piece/month ② Between 0.09% piece/month and 0.06% piece/month ① Less than 0.05% piece/month
		7	Effect of equipment/machine on quality of end items	5	4	2	1	⑤④ Decisively affects end items' quality ② Somewhat affects end items' quality ① No particular effect on end items' quality

○ Remarks:	Total points	A B C

Ranking criteria
(Judged on basis of points
and remarks)

A......... More than 23 points
B......... From 22 to 16 points
C......... Less than 16 points

FIGURE 3.7

63

that everyone have accurate results from the type of analyses described before. With regard to accurate understanding of a breakdown, it is important to grasp the modality of the breakdown mechanism. Thus, if everyone involved in the eradication of breakdowns can provide tabulated data showing the mechanical positions and causes of the breakdowns, and can discuss these relationships, the level of usefulness of the data is very high. In this regard, Figure 3.8 presents a comprehensive table of situational analyses of sudden breakdowns.

Furthermore, analysis of a breakdown must be carried out in relation to the current productive maintenance management system. This means a review of the existing various sets of standards. Such a review should be carried out by first determining who should be conducting the review and on what time schedule; the review should then be followed up at the PM collaboration meetings.

3. CRITICAL COMPONENT ANALYSIS (CCA)

With regard to critical equipment, an estimate of the number of breakdowns in their critical components should be made in advance, to cope better with the situation. Such an analysis is called a critical component analysis (CCA). The important aspects of this analysis should be summarized in a table to facilitate its use in the prioritized management system. These aspects are (1) the inventory of the critical components, (2) their predictable breakdown modes, (3) the predictable causes of breakdown, (4) possibilities and methods of anticipating breakdowns, (5) estimation of time needed for recovery from each breakdown, (6) effects on secondary breakdowns, (7) availability and costs of carrying spare parts, (8) estimation of procurement lead times, (9) productive maintenance plans and standards for preventing equipment deterioration, and (10) plans for improvements.

A considerable amount of time is necessary to complete these analyses, since they require significantly detailed engineering investigations. A useful approach to this phase of the activities is first to focus on the production lines that have a high load level. Then the priority is to take on the challenge of zero breakdown in these areas and spend as much time as needed until they are thoroughly tuned. Once that is accomplished, then the perfected lines can be expanded gradually. Figure 3.9 shows an example of this strategy.

The following sections will discuss further breakdown-prevention activities, dealing with (1) reduction of minor short stoppages, (2) workers' self-initiated maintenance roles, (3) planning and management of productive maintenance, and (4) quality assurance.

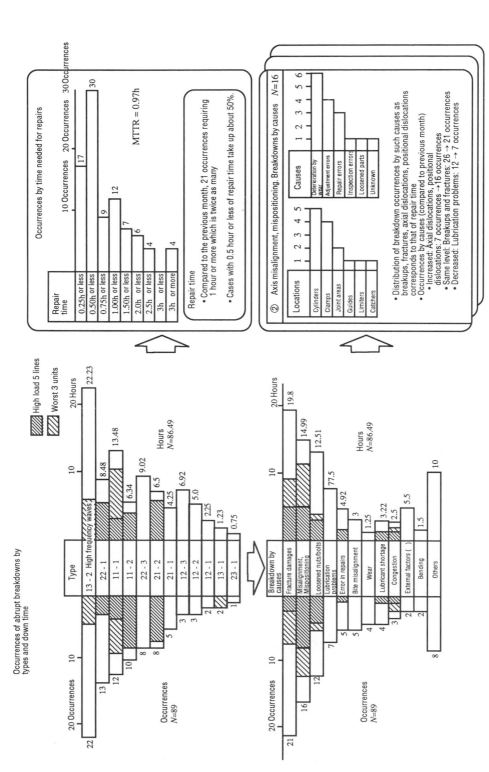

FIGURE 3.8 Overview Table of Analysis of Abrupt Breakdowns in Plant for a Given Month

65

Critical component analysis (CCA)

Plan view

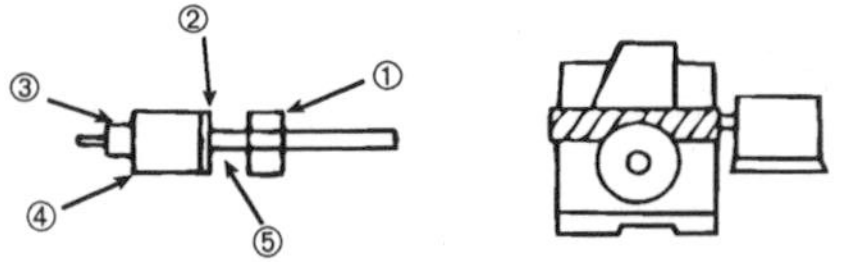
①
②
③
④
⑤

No.	Critical components	Breakdown mode	Breakdown cause	Predictability	Recovery time	Presence of secondary breakdowns	Spare parts: Avail-ability	Spare parts: Lead time	Emergency measure: Possibility	Emergency measure: Time needed	Evaluation
1	Bearings	Burnout, fracture	Dried-out lubricant, vibration	Vibrometer	2^H	Yes	Yes	7^H			A
2	Cover	Corrosion	Liquid infiltration	Visual check	3^H	Yes	No		Yes	3^H	C
3	Joints	Loosening	Design related								
4	Worm gear's shaft	Break up	Overload-caused fatigue								
5	Key	Break up	Joining problem								
6	Wheel	Total wearout breakdown	Overload-related dried-out lubricant								
7											

Critical component maintenance standards

No.	Critical components	Maintenance mode	Standards	Method	Cycle	Assigned to:	Treatment
1	Bearings	Crack					
2	Cover	Corrosion					
3	Joints	Abnormal noise					
4	Worm gear's shaft						
5							
6							

Summary of critical components and improvement plan

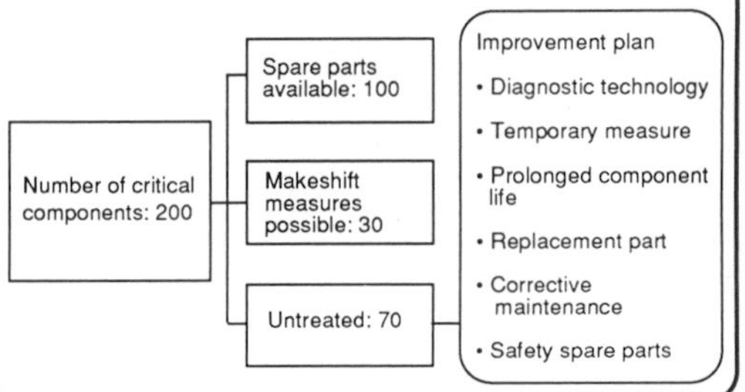
Number of critical components: 200
Spare parts available: 100
Makeshift measures possible: 30
Untreated: 70
Improvement plan
• Diagnostic technology
• Temporary measure
• Prolonged component life
• Replacement part
• Corrective maintenance
• Safety spare parts

FIGURE 3.9

3.2.4 Analysis of Assembly Line Short Stoppages from Minor Causes and Their Remedies

In the assembly line environment, production interferences caused by brief and minor stoppages are often of greater significance than is usually recognized. These brief stoppages or irregularities that disturb production cycle times—because of the brief duration of the stoppages—are often too difficult to track and count accurately, or to record their specific causes, whose identification is crucial.

We have investigated the number of occurrences of brief stoppages in a plant comprised mainly of presses and welding assembly machinery; Figure 3.10 illustrates the results. Depending upon the

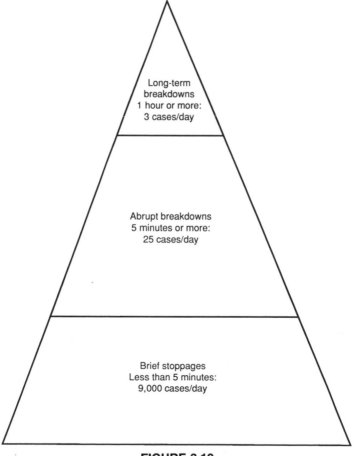

FIGURE 3.10

production lines, there is a varying frequency level of brief stoppages; similarities, however, are often found among them. Brief stoppages are defined as those stoppages that disturb production cycles, i.e., the standard operational pace. For example, such stoppages may take the form of a machine whose system for automatic impact unloading of parts will not eject finished parts unless it is manually tapped, or a machine whose fixtures do not accept a workpiece unless the piece is manually inserted. The workers at such workstations could become accustomed to the idea that these ejection devices and fixtures are designed to function this way, and thus would learn to cope with these conditions without complaining.

Other examples of brief stoppages include excessive non-work-related conversation among workers; unnecessarily pushing a switch button twice, bending to pick up a fallen part, manual dislodging of a workpiece stuck in a parts conveyance duct, manual removal of scraps beyond what is necessary, idling caused by the delayed arrival of components in short supply, occasional quality inspection, stoppages caused by scrapped parts, changeover and adjustments of lathe bits and fixtures, changeover of part bins, and so on. The floor managers, supervisors, or operators themselves usually respond effectively to such brief stoppages without calling on the maintenance crew. The maintenance personnel, therefore, tend to lose their visibility and are often hard to identify.

In terms of duration, a brief stoppage refers to one lasting from one second to five minutes. Reduction in the occurrence of brief stoppages improves equipment utilization. It also, however, raises numerous technology-related issues in the context of the current trend toward increased automation.

Brief stoppages not only interfere with plant productivity, but also contribute to a subconscious mistrust of mechanized equipment on the part of the operators themselves. This could lead to underachievement and the attribution of problems in quality to the machinery, causing production shortages and equipment problems. In other words, brief stoppages contribute to a lowered morale of the operators. In many cases, however, as noted before, brief stoppages occur due to simple causes. Therefore, when PM is to be implemented on a companywide scale, the subject of brief stoppages can be naturally and spontaneously discussed among operators and supervisors in group activity meetings. The results to be derived from such activities can be significant, helping to raise the morale of workers with regard to their participation in TPM.

The reader is encouraged to investigate the process of how measures are taken to eradicate brief stoppages, for which Figure 3.11 is

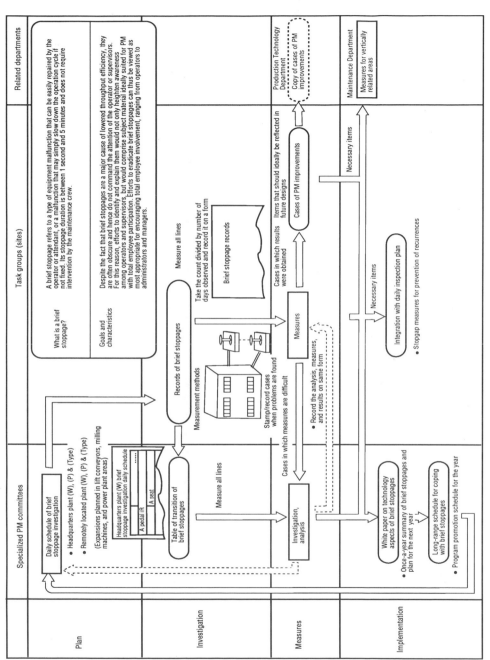

FIGURE 3.11 Mechanics of Implementing Measures Against Brief Stoppages

provided as a reference. It should be noted that since we are dealing with breakdowns that are obscure and minor in nature, it takes some imagination to identify and explain them. Counters should be set up at locations where frequent breakdowns are observed. They should be tallied once a day, and the totals recorded in a designated form. In the case of the plant used in the figure, the activities were first recorded on video, which was reviewed by each personnel group during their lunch breaks to apprise them of the situation. Then a survey was conducted on the frequency of stoppage occurrences. This level of tenacious preparation is necessary for a PM program to succeed.

3.2.5 Record Keeping and Record Use for Productive Maintenance

—KEY POINTS IN MANAGING TECHNICAL INFORMATION ON
   MAINTENANCE: TABULATED RECORDS OF MTBF ANALYSIS—

1. PROBLEMS RELATED TO OBTAINING MAINTENANCE RECORDS AND THEIR USE

The method of obtaining and using maintenance records is directly related to a "snapshot" evaluation of the management level of maintenance activities and of the skill level of maintenance technicians. Maintenance records and their use comprise the foundation of maintenance management, and are indispensable to maintenance engineering activities. When we visit a site to consult on plant maintenance, maintenance records and the reality of how they are used are the first elements to be evaluated.

   In some plants, a PM program may have been in place for a considerable period of time where maintenance cards are kept based on maintenance records and tuning and repair order stubs. There may be a cabinet in the corner of an office in which racks of these maintenance cards may be neatly stored in alphabetical order. But when the contents of the record cards are examined in detail, there is often little significant information on the cards. We cannot help but suspect that, regardless of how many years such cards have been kept, it may be impossible to use them for either maintenance management or engineering activities. Worse yet, in some cases, neither the maintenance crew members nor the maintenance engineers have enough time to record data on them, and secretarial personnel are requested simply to transcribe the contents of repair and tuneup stubs onto the cards. In such a plant, the cards are hardly used by managers, by supervisors, or by the maintenance crew in their maintenance management or engineering activities.

Why does this type of situation—with some variations—arise even in plants with significant track records of maintenance management? Usually, it occurs because there is a lack of clearly defined common understanding among various department personnel involved in PM regarding the use of raw maintenance data in future maintenance management and equipment engineering activities. In other words, the ways of utilizing the data are not the results of their own thoughts and ideas, but are instead mere imitations of other businesses. Even though these people may attend seminars, attempt to adopt whole systems of advanced journal entry methods, and mimic the maintenance record-keeping methods of other businesses, these practices may not necessarily be appropriate to the level of management realities in their own company. Moreover, if these people do not opt to become involved in some ongoing analysis, but rather try to introduce new techniques through the use of existing staff people, the chances of failure are very high.

Specifically, all personnel should also recognize the value of certain raw data and create a record form that is well suited to the characteristics and level of the company, regardless of how commonplace the records may appear. Furthermore, the data should be used by linking the information to management and engineering action. In this way, levels can be raised one step at a time.

When maintenance engineers are engaged in management or engineering activities, their activities must be based on maintenance data. Activities that are not backed up by accurate data or other factual information may be derived instead from the experience of seasoned veterans, but these actions cannot be called engineering activities. A maintenance engineering activity may be defined as one that, through the use of diagrams and data, deals with a clear causal relationship or attempts to clarify the causal relationship between a phenomenon and its causes, or a cause and its effect on a machine. It refers to an activity that always deals scientifically with a fact and the data pertaining to it.

2. THE VALUE AND GOALS OF MAINTAINING AND COLLECTING MAINTENANCE RECORDS

As equipment investments increase, equipment maintenance tasks expand proportionately with regard to required skills and technology. The number and volume of these tasks, moreover, also increase dramatically.

In order to accomplish these tasks effectively with a limited maintenance staff and to increase equipment productivity, the following steps are necessary: (1) reduce to zero the occurrences of unscheduled maintenance work within the facility; (2) devise measures, from a variety of engineering perspectives, that afford a maximum necessary time cycle needed to carry out planned maintenance work (inspection, lubrication, periodical maintenance, etc.); and (3) reduce the number of operational steps required for each maintenance event. These activities may be defined as equipment study for improvement of equipment reliability.

On the other hand, if it is uneconomical to expend efforts to reduce to zero the number of occurrences of maintenance work within the facility, it is necessary to investigate efficient ways of processing equipment work requirements. This type of investigation may be defined as equipment maintainability research. It includes investigation into methods of replacing groups of parts in the least possible time and into means for improving the repair facilities in general.

If these two studies are thoroughly carried out, a PM system will result in a marked level of improvement and efficiency. It is necessary to understand maintenance activities that arise in actual situations, analyze their problems, define specific areas, and make proposals for improvements aimed at the following: (1) to select the appropriate themes of the equipment study, (2) to improve the reliability of equipment, and (3) to carry out activities to improve the equipment's maintainability.

This means simply that the basic information required for improvement is nothing other than the maintenance record itself. The purpose of improvement is always to reduce the number of tasks and to increase efficiency. The maintenance record is both convenient and easy to use, if its compilation is based on each unit of equipment and it contains a table analyzing the characteristics of the maintenance tasks that were actually undertaken.

The analysis of the characteristics of various maintenance tasks arising from equipment (repair of unscheduled breakdowns, planned maintenance, improvement maintenance, planned inspection, lubrication, adjustments, cleaning, and similar activities) shall be defined here as the MTBF Analysis. MTBF is an acronym for mean time between failures, and it stands for the average time intervals between breakdowns of equipment or components. Since it is used for analyzing the intervals between failures, it connotes the efforts to determine an adequate maintenance period.

3. GOALS AND APPLICATION OF MTBF ANALYSIS (ANALYSIS OF MAINTENANCE RECORDS)

(1) Goals of MTBF Analysis

By analyzing the frequency of occurrences, engineering modifications may be carried out to respond to unexpected failures and prolong the useful life of parts and components, with an emphasis on those that require frequent maintenance.

(i) Selection of areas for improvement and reduction of maintenance requirements.

(ii) Estimation of life span of parts and study of an optimized repair plan. The estimation of the life span of parts is made on the basis of scattered distribution of times between failures as well as on pattern characteristics. The soundness of repair methods should be examined from the findings of the first breakdown after a planned maintenance. Based on analysis of these data, an optimized repair plan should be made for a particular machine or a group of machines.

(iii) Selection of points to be inspected and determination and modification of inspection standards. The key points and areas of key equipment, or the areas in which relatively frequent failures occur, should be studied as a subject of key maintenance (preventive maintenance).

(iv) Selection of in-house or outside maintenance. A question may arise on the division of maintenance work assignments between in-house and outside maintenance options. In terms of securing repair quality and increasing equipment reliability, it is important to divide maintenance work assignments according to the equipment and type of activities, based on the assessment of either in-house or outside maintenance skill levels, quality requirements, and the volume of work that is to be carried out.

(v) Determination of standards for spare parts—basic update data. Identification of the required quantity of spare machine parts and electrical components to be regularly stocked must be based on the frequency of use pattern as well as the lead time needed to procure each part. The recurring problem in determining spare parts standards results from the number of operational steps required to carry out the investigation. This, in turn, is caused by the lack of a clear and sys-

tematic way to examine and modify each stock item from maintenance records on a day-to-day basis.

If, however, the appropriateness of regular stocking is examined for each component based on the MTBF Analysis Table each time a failure occurs, then spare parts inventory management can be accomplished accurately and easily.

(vi) Key points for improving repairs and tuning methods. Some equipment requires a great amount of time to repair and tune, as well as to replace parts. To improve the equipment utilization rate, it is necessary to reduce the time needed for such operations on this equipment.

To this end, maintenance methods must be improved and the bulk parts replacement method investigated. The MTBF Analysis Table will provide the basic information needed to select the agenda to be investigated by suggesting a ranking order of priorities.

(vii) Maintenance work selection for establishing estimated time standards and time standards study. Before selecting the kind of maintenance work to be undertaken so as to establish estimated time standards for repairs and tuning, the following should be considered: frequency of occurrence, fluctuations in actual time needed to complete the work, and the characteristics of the work to be carried out. For this purpose, the use of the MTBF Analysis Table is indispensable.

(viii) Selection of key equipment and components for diagramming. The organization of information about equipment components to be improved in the MTBF Analysis Table, revisions to diagrams resulting from depreciation and deterioration, and drawings needed for making spare parts should all be as clear as possible. Ideally, these diagrams should always be used as guides for maintenance activities.

(ix) Assigning responsibility for establishing and modifying operational standards and maintenance work.

(x) Engineering data for reliability and maintainability design. The previous section dealt with the PM activities system and the key points of its operation. As noted before, engineering modifications derived from daily PM activities of maintenance engineers are often not utilized in the design for equipment reliability and maintainability. This occurs as a result of the maintenance crews' failure to provide engineering information about equipment reliability and maintainability or to organize the information so that design engineers can use it in their

maintenance activities. Furthermore, design engineers do not try to standardize engineering data so as to increase the group's collective skill level, nor do they constantly gather maintenance information and edit it so that it can be used for determining design standards for equipment reliability and maintainability. In short, the problem is the gap between the activities of the design and layout engineers and those of the maintenance engineers.

It is important for maintenance engineers to gather engineering information aimed at reliability and maintainability designs. This information should be based on the MTBF Analysis Table and communicated to the design department. Such efforts will also prove useful for organizing engineering tasks.

(2) Data to Increase Understanding of the Relationship Between Equipment and Product Quality

It is difficult for the maintenance department to grasp the linkage between maintenance activities and product quality. When repairing equipment breakdowns, the focus of the attention is bound to be on restoring the lost functions. As a result, there are cases in which quality improvement resulting from maintenance work may not be recognized. But since there is an important link between equipment and product quality, it is important to relate product quality, equipment, and maintenance activities, to organize and consolidate their key points, and to create an agenda of actions to be taken.

(3) Development of Diagnostic Technologies

PM emphasizes anticipatory maintenance, focusing on diagnostic technologies in equipment. However, it is often difficult to define specific issues in this area. A pragmatic approach is the use of the MTBF Analysis Table to find topics of development for diagnostic technologies.

(4) Creation of Educational Materials About Equipment

A major PM theme is the development of personnel who are thoroughly familiar with their plant's equipment. This raises the question of what specific points should be taught and how they should be taught. In general, engineering education does not go beyond the compilation of textbook knowledge, which inevitably fails to relate to one's own company and the particular problems of the company's workplace. To bridge this gap, it is necessary to concentrate on pragmatic aspects by conducting an MTBF analysis and also to teach about the actual equipment's functions and design structures, including its weaknesses.

Ideally, to enable the maintenance staff to work with diagrams, a combined use of the MTBF Analysis Table and related drawings should be practiced. All the related drawings should be kept in one place, in a rack near to where the MTBF Analysis Table is posted. Whenever entries are made on the MTBF Analysis Table, the related drawings are displayed for discussion and, if necessary, remarks are written in red ink in appropriate places on the drawing. Thus an "engineering" mind is developed, one that is accustomed to working with drawings. Skills are also enhanced for easily translating maintenance ideas into simple diagrams. Maintenance crews that do not fully acquire the habit of working with diagrams would tend to forget how to analyze or make them, even though they may have been exposed to such drawings in a classroom environment. Unless maintenance crews are trained in depth to become maintenance engineers, their effectiveness will not markedly improve, no matter how long they may work at it.

(5) Understanding and Studying the Concept of Life Cycle Cost

The MTBF Analysis Table shows a panoramic view of data in a series over a relatively long period of time. It is simple, therefore, to see the occurrences of maintenance activities, costs, spare part replacements, and other losses as a life cycle cost picture. Furthermore, the table allows review of the measures taken for each maintenance case, providing answers to the question of whether the measure taken should have been one of repair or replacement, as well as showing the life history data of each piece of equipment.

As noted earlier, the MTBF Analysis Table is only one method for keeping and using maintenance records. These records, however, have an extremely high value, insofar as they point to guidelines for maintenance and related engineering activities. It is not an exaggeration to say that PM activities begin with the making of maintenance records and end with accurate record keeping and utilization of maintenance data.

4. CREATING AN MTBF ANALYSIS TABLE

—CHARACTERISTICS AND BENEFITS OF TABULATED MTBF
ANALYSIS RECORDS—

Figure 3.12 is an example of an MTBF Analysis Table. In addition to the tabulation technique, several mapping techniques are shown, such

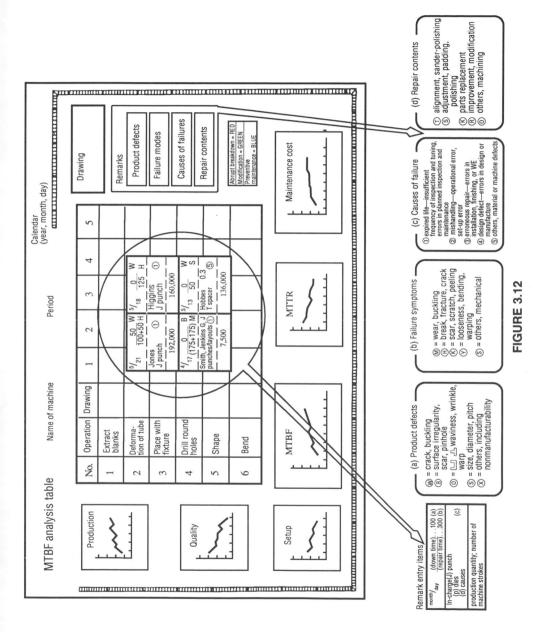

FIGURE 3.12

as those using product diagrams and equipment plan views. As long as the purpose of MTBF analysis is highlighted, any one of these methods should serve to define and utilize maintenance records.

(1) Characteristics of the MTBF Analysis Table

An explanation of the characteristics of the MTBF Analysis Record Table is appropriate here, since the characteristics define the key points in creating such a table.

(i) The data in the table should be accessible at a single glance, and should therefore be condensed into one page. In the past, maintenance data were usually recorded in a one-case-per-page fashion, thus complicating investigation into their past history. If the data are compiled into a single table and are not scattered among numerous pages, an overall view of the production lines and equipment will be quickly apparent.

(ii) Data for a certain time period can be viewed within the time frame of its actual development. In order to analyze a breakdown, the data should allow us to see the history of past breakdowns for a particular component or part, the type of remedial measures then taken, their effect, and the subsequent frequencies and intervals between breakdowns. For the data to be used in this way, it is important that the data-gathering method should be time-series oriented and continuous. On the other hand, the method of gathering data at fixed time periods varies, depending upon the characteristics of the equipment. As long as this concept of data gathering at artificially determined intervals is defined in a manner conducive to the equipment's life cycle maintenance, however, it can serve as a life history log of the equipment.

(iii) Maintenance record keeping and analysis goals can be met simultaneously. Maintenance data are characterized by information about lengthy breakdown intervals for any particular part or component, since cases of recurrent breakdowns are rather rare. Nevertheless, when a breakdown does occur, it is desirable that remedial measures be based on analysis of past events. Even if a monthly maintenance report recording the breakdown is published, it would not serve the same purpose, since it would not contain information about past breakdowns.

Thus, for maintenance work it is important to have data available that serve for both record keeping and analysis. If such data provide an overview capability no matter how large a chart might be required to display it, and if this chart is hung on a wall of the shop floor or wherever maintenance activities are actually taking place, it would serve its purpose even better.

(iv) With proper format design, more information can be condensed on a single chart. A variety of information and data is generated on the shop floor, but most is discarded after being gathered. Maintenance data tend to be the type most often retained, and they convey most of the information available about the shop floor. In addition to their use as the primary information source on breakdowns and maintenance, if information on quality, safety, costs, and so on, are also incorporated, then maintenance data records could be changed into effective engineering data for manufacturing, designing, and engineering.

(v) Understanding the maintenance focus can be facilitated. Maintenance records usually do not clearly indicate where the maintenance was concentrated. But if diagrams, symbols, and other markings such as color distinctions are used within the MTBF Analysis Table to highlight problem areas where frequent or major breakdowns occur, it is possible to clarify and underscore the areas where action should be focused.

(vi) The relationship between breakdowns and maintenance can be understood. Knowledge of a breakdown's cyclical occurrence can be useful in dealing with equipment reliability. Even with this knowledge, the maintenance crew needs to determine what type of action to take, how much time should be allotted for replacement maintenance to be done, and when inspection should be undertaken for each maintenance task. The results of maintenance activities should all be noted in the maintenance calendar. Using an MTBF Analysis Table, the maintenance perspective links together breakdowns and associated maintenance activities, and can be related to both the maintenance records and calendar.

(vii) Remedial measures and their effects can be understood. Measures taken for each breakdown relate to the life span of the equipment, and their effects cannot be immediately assessed; it may require months or even years for the effects to be known. Therefore, it is difficult to assess the effectiveness of measures taken. A maintenance record usually states what was done, and there the record ends. On the other hand, an MTBF Analysis Table can indicate the circumstances surrounding a certain specific measure and evaluate its effect.

The history and weaknesses of a machine cannot be understood by observation alone. An MTBF Analysis Table, however, can be a very effective tool both for understanding a machine's past and for enhancing follow-through capabilities.

(viii) Overall understanding is increased. Visual recognition is crucial for action. But conventional maintenance data tend to be too specific and individualized, obscuring their overall focus. On the other hand, administrators cannot afford to spend large amounts of time looking through details. If a management focus is clearly presented on a single sheet of paper, it becomes easier for managers to relate to it and, thus, see the focus and understand it.

(ix) Record entries are made simpler for everyone. Conventional maintenance data are usually recorded only by the maintenance crew. Record entries often omit important items and are too cryptic for others to understand. The first PM step was taken as an effort to keep maintenance records correctly. By subsequent, ingenious methods that developed in data notation—such as the use of pictographs, symbols, and color-coded markings—it has become possible today for anyone to make the record entry easily, thus facilitating the gathering of good data. These innovations have helped promulgate the importance of maintenance among production department personnel as well as heightening the maintenance crew's awareness of the importance of these methods. The MTBF Analysis Table can itself function as the educational material for these purposes.

(x) The daily raw data become the management data. Raw, unprocessed data are of little use. However, the entire cycle of processing and analyzing raw data before possible action may take too long. It is important, therefore, to improve the method of gathering data in such a way that raw maintenance data require only daily gathering, and so that the analysis is done as an intrinsic part of the method.

MTBF (mean time between failures) = mean failure interval

Computational example:

$$\text{MTBF} = \frac{\text{Operational period}}{\text{Failure occurrences}} = \frac{12 \text{ months}}{3 \text{ times}} = 4 \text{ months}$$

In addition to the mean value, sometimes a three-point estimation technique may be used.

While considering the objectives of the table:

1) Determine the object and scope of the analysis table:
   Whether or not to focus by production line, similar machines, or components.

2) Prepare the form: A large form is preferable for initial learning purposes.

3) Classification and segmentation of the objects of analysis: Retain tabular overview characteristics of the table with capability to compare frequency of occurrences, etc.

4) Determine format of the analysis table: Be creative. Allow for inclusion of past data, as far back as six months to one year. Use of diagrams should make them more understandable. Try to design a one-page table. Incorporate spare parts information as well.

5) Determine contents to be recorded: Date of occurrences, number of units manufactured, name of the equipment involved, names of parts or components involved, name of person(s) who dealt with the event(s), failure modes, causes, measures taken, type of maintenance performed (abrupt breakdown measure, prevention, improvement, etc.), number of operations involved in maintenance, maintenance cost, etc.

6) Consider method for recording the contents: Symbolization, use of pictographs, color-coding and differentiation, facilitating understanding, measures taken for upkeep efficiency.

7) Record in the analysis table: Record whenever problem occurs.

FIGURE 3.13 Procedure for Creating an MTBF Analysis Table

(2) The Procedure and Three Patterns for Building the MTBF Analysis Table

(i) The procedural steps for building the MTBF Analysis Table are shown in Figure 3.13.

(ii) Understanding the objective and scope of building the MTBF Analysis Table. The MTBF Analysis Table belongs to one of the three patterns shown in Figure 3.14.

Pattern (i) is characterized by an approach that emphasizes high-priority equipment and high-priority production lines. This approach is often used as a model when PM is newly introduced. It calls for determining which equipment and lines within a plant should be given high management priorities. Thus, it is suited to a management approach that calls for thorough analysis of the prioritized equipment. Just as for plant activities, a selective prioritization plan should also be designed for plant equipment. Maintenance carried out on the basis of such a plan will be effective. Newly installed and problematic equipment should always be managed with special attention paid to the priority rules. The problem with this approach, however, is that it raises the question as to what extent lower-priority equipment should be maintained, and it fails to answer this question clearly.

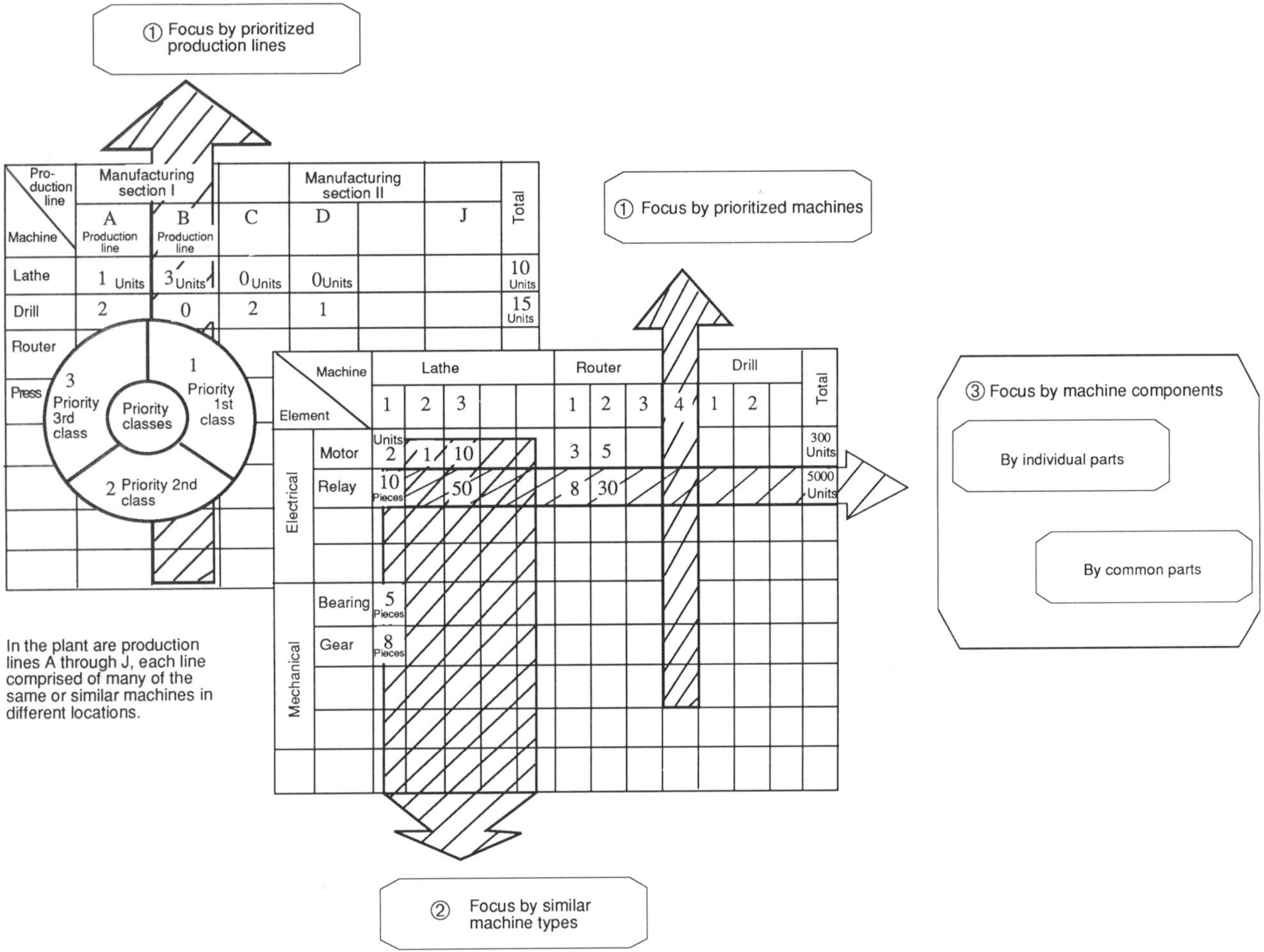

FIGURE 3.14 Model of an MTBF Analysis Table

In the plant are production lines A through J, each line comprised of many of the same or similar machines in different locations.

Pattern (ii) is an approach based on an equipment-type classification. According to this approach, maintenance is conducted by analyzing the similarities between the actual equipment and the equipment as described in its classification catalogue. Even though the plant may not have a full-fledged high-priority production line, the same or similar equipment may exist to some degree, and so comprise a production line. Observing breakdowns of these related machines could point to common or similar problems. This approach consolidates data from similar machines, which may vary in number from several to dozens of units, and records their breakdowns on one page of their MTBF Analysis Table. Records of all the plant equipment can thus be kept on a very small number of MTBF Analysis Tables. Cases may occur in which the aggregate breakdown occurrences may be few and far between, and they are more easily understood if similar machines, fixtures, and molds are reviewed collectively. In such cases, this approach is very effective.

Pattern (iii) is an approach based on a type classification of machine elements, such as equipment parts or components. If machines are analyzed, they are found to consist of similar parts and components, such as electrical and mechanical parts, but observing these parts alone would not reveal to what kind of machine they belong. Many parts with specific functions are combined in a machine. Maintenance or specialized engineering jobs are usually divided into such specialized areas as electrical and mechanical. It follows that analysis of a breakdown would be based on specific parts or components. It is therefore relevant to analyze them as common parts or components. It should also be noted that, even when machines are thought of as "high priority," we are actually paying attention to high-priority parts and components within those machines. In fact, this mode of thought is necessary for the analysis to have any practical effect. An ultimate MTBF analysis, therefore, should call for an analysis by such elements.

Three different patterns and approaches were described. The closer one gets to the third pattern, which is based on the classification of equipment elements, the greater the details and more unwieldy the classification, thereby obscuring the overall picture. On the other hand, an opposite approach such as described in the first pattern—which does not deal with the function or design characteristics of the machinery, but concentrates only on the classification of priorities—would be so diffuse as to again require analyzing more detailed, important maintenance areas. Thus, each of the three approaches has strengths and weaknesses. An approach should be chosen only after careful thought has been given to the characteristics, goals, and level of sophistication of the company's personnel.

(3) Noteworthy Points on Using the MTBF Analysis Table

An MTBF Analysis Table is only one way to keep maintenance records, and is by no means a panacea. Unless the people who are directly involved design a format best suited for them and easiest for use by them, the table will become just an idle ornament. In this regard, a number of important points should be considered.

1. After the MTBF Analysis Table format is determined, it should be created on the basis of past data and should clarify the real-world picture of the equipment. For equipment with frequent breakdowns, six months of past data should be sufficient. Even with records where data are missing, if data from the past three years or so are used, an overall profile of the equipment can emerge. It is important not only to retain past details, but to focus on a well-organized view of the problems. As measures are taken to cope with newly arising problems, records should be kept daily and consistently.
2. As the words "mean interval time between failures" indicate, an MTBF analysis offers a statistical view, because it focuses on cyclical breakdown periods. In reality, failures include both those resulting from human error and physical breakdowns. Moreover, some parts may not have failed or broken down for several years. There may be a scarcity of data for other parts or components, or uneven intervals for those data that do exist. A failure analysis based on an MTBF Analysis Table should not just present a statistical view, but should also deal with engineering aspects, i.e., questions of what happened, and what should be done. In an MTBF Analysis Table, therefore, it is important to clarify the type of failures that require engineering solutions, such as fatal and recurring breakdowns.
3. As records are kept in an MTBF Analysis Table, failures and repairs for frequently failing and high-priority equipment are entered. As preventive and improvement maintenance is carried out for this equipment, entries of unscheduled repair maintenance will begin to disappear from the MTBF Analysis Table, and preventive and improvement activity entries will increase until they dominate the table. When this happens, the MTBF Analysis Table will be identical to what is noted in a planned maintenance schedule, thus eliminating the need for keeping the MTBF Analysis Table. When this stage

has been reached, unless the format of the MTBF Analysis Table is modified to incorporate similar production lines, similar equipment, and similar parts and components, the usefulness of the MTBF Analysis Table will cease. Conversely, such a modification will enable it to continue as an effective record-keeping method.

4. Once a broken part or component is repaired, some time will usually pass before that part or component breaks down again. Thus, no problems will arise even if maintenance records are not kept for a part or component while it is functioning properly. However, if maintenance records are set aside and not noted on the MTBF Table, record entries are certain to become more difficult later on. This could lead to a haphazardly maintained record table. Since effective maintenance depends upon planning for future measures based on review of past breakdowns, it must become a matter of routine that the record entry is made each time a repair is made. Moreover, the MTBF Analysis Table should always be used after a repair is finished, or in a maintenance planning and reviewing session. If this record table remains sparkling clean, without smears or fingerprints, it is a sure sign that the table has lost its effectiveness.

3.3 ANALYSIS OF THE RELATIONSHIP BETWEEN QUALITY (Q) AND EQUIPMENT

—CREATION OF PRODUCTION LINES GUARANTEEING QUALITY
   THROUGH RELIABLE EQUIPMENT—

Equipment breakdowns are generally divided into abrupt breakdowns, functional deterioration types, and, especially critical, qualitative deterioration breakdowns. Many people perceive the goal of PM activities as prevention of only physical breakdowns. Because of this tendency, evaluation of PM activities is likely to focus on factors such as the frequency of failures and rate of physical endurance of the equipment. PM is itself often understood in the narrow sense of literally "preventive maintenance," connoting inspection, lubrication, cleaning, and repair. Thus the tendency is to limit the meaning of PM to the maintenance crew's traditional range of activities.

PM, however, is a much broader technology that takes on the challenge of improving factory activity outputs: P, Q, C, D, S, and M. As a matter of course, therefore, maintenance technicians should thoroughly understand production-related technologies just as production technicians should understand maintenance-related technologies, so that they can clarify the interrelationships between aberrations in product quality and the conditions of production equipment and equipment deterioration, and rigorously promote activities to reduce the scrap rate and work-in-process inventory.

In addition to the maintenance and production technicians, the quality control staff and the production engineering staff should also carefully analyze the scrap statistics and management charts. These technicians and staff should then understand that the measures taken to reduce the scrap rate can always produce effective results when linked to manufacturing technologies and their applications in equipment or equipment technology. When we refer to quality maintenance engineers, we really mean those people who have the ability to solve problems and not those who draft management charts based on current situations. In this context, the required educational preparation and experience become apparent. It is necessary to define clearly the capabilities expected from quality engineers and to train them accordingly.

In addition to quality failures caused by equipment problems, a significant number also results from either human errors during the manufacturing process or from inferior materials. Among the prize-winning companies in Japan, the number shipping inferior lots has significantly decreased. Moreover, the chief executives of these firms have recognized that they may personally contribute a great deal to total employee participation programs to promote PM activities and thus reduce costs. Among these companies, TPM is now highly valued as a management method that closely reflects the realities of the actual work environment and is directly related to the profitability of their enterprises.

3.3.1 Goals and Values of Quality Maintenance Activities

The concept of quality control is based on three principles: "nothing inferior should be received," "nothing inferior should be produced," and "nothing inferior should be shipped." Japanese products exported to the world marketplace that account for the growth of Japanese industry are characterized by stable quality levels. Until recently, many people were assigned to inspect the finished goods during the final inspection phase in order to prevent complaints about products. With such an inspection regarded as indispensable, both inspection methods and machine fixtures were changed to reduce the number of inspection steps. Efforts were also invested in developing automated inspection equipment. But all the foregoing are intended to remove inferior quality products; they are not actions aimed at lowering the scrap or waste rate itself.

In the next stage, a method (known as preautomation) was devised in Japan using sensors to identify and stop machines producing inferior quality output. Methods were also investigated aimed at preventing production of inferior products before output. In light of current trends toward unattended equipment operations, this investigation should be intensified. In the context of PM, the relationship between key points in producing equipment-dependent quality goods and equipment failure modes should be clarified. This relationship is used as the basis for defining quality maintenance activities involving all employees, including the quality maintenance staff, production and maintenance crews, and all other personnel, so that the prevention of quality failure becomes a total collective activity within the company.

Since its inception, PM has sought to implement a planned maintenance program to act before equipment failures occur, taking on the challenge of zero breakdown. To this end, attention has been paid to the intervals between failures, and optimized maintenance cycles were investigated. Although the conventional quality maintenance approach has also taken similar measures against problems of inferior quality, it has lacked conceptual rigor when compared to PM. In other words, it seems that conventional approaches to quality maintenance tend to have a less stringent focus with regard to on-site activities for preventing problems of inferior quality.

The current widely accepted quality management approach is to understand critical areas by using the Pareto analysis technique for collecting statistical data on various failure modes. Following this, failure modes are classified by their components—people, equipment, raw materials, methods, and so forth—based on diagrams illustrating key factors, so as to highlight and correct the problems. Another noteworthy aspect deals with the problems of quality improvement efforts through total employee involvement in quality group activities. This approach also advocates the frequent use of maintenance charts at plant sites.

However, the maintenance charts in this context are intended to predict symptoms of failure modes in terms of products. The use of maintenance charts to describe causes of or factors contributing to failures is neither sufficient nor satisfactory. That is to say, the analyses and responses derived from these charts tend to focus on modes of failures that actually occurred. This approach, however, does include some preventive measures, insofar as problems related to equipment, machine fixtures, and molds are regarded as part of the maintenance domain. Nevertheless, it does not give maintenance department personnel a strong sense of a commitment to the idea that "there are many areas in which we must guarantee quality." In this approach, therefore, there are insufficient preemptive activities that are structured to ensure output of quality products.

In PM, such activities are called quality maintenance. The PM approach aggressively promotes "activities to prevent quality failures" by thorough and relentless pursuit of preventive maintenance for quality assurance, focusing on everything related to maintaining and building quality, such as the equipment, molds, dies, fixtures, error-arresting devices, and measuring instruments. It also focuses on preventing failures in those tools and instruments that measure and evaluate quality.

Quality control techniques such as the widely used Fault Tree Analysis (FTA) technique to trace an equipment failure down to its exact parts or components, or the prediction of impending failures by the use of maintenance and transition charts, should all be studied for possible broader application. Just as preventive maintenance for the total elimination of equipment stoppage type failures is accomplished by tracking the failure sources down to the equipment component levels, so, too, efforts to maintain the upper hand in preventive maintenance for quality assurance would require preventive maintenance of all equipment components that have anything to do with quality. Otherwise, successful PM is not possible. In short, it is important to establish the concept of quality preventive maintenance (see Figure 3.15).

Concept of Preventive Maintenance for Quality

In the process of successfully developing a product, its manufacturing process plans are evaluated, and they, in turn, are translated into the key production means such as the equipment, molds, and fixtures used in the new product's manufacture.

The quality of a product is determined by both the equipment engineering and production technologies, and by the constant efforts of the production staff. Although TPM advocates an equipment-oriented management philosophy, focusing on the maximum use of equipment, in fabrication and assembly type manufacturing environments it emphasizes the idea of "building the quality into the product through quality-oriented equipment." The term preventive maintenance for quality was born in this context. It recognizes the need for thorough quality management throughout the entire life cycle of the means of production.

It focuses on means of production such as the equipment, molds, dies, fixtures, gauges, automated quality inspection machines, reliability testing equipment, industrial instruments, counters and measurement instruments, error-prevention devices, cutting tools, and so on, throughout their life cycles: from initial development and design, to manufacture, purchasing, use, and depreciation.

Unless the sources of quality failures, as mentioned before, are traced and identified in a proactive manner, it is not possible to assure the perfect quality of products. In contrast to preventive maintenance for equipment, this is called preventive maintenance for product quality. For this purpose, the components of equipment that are related to product quality are clearly identified as quality components, and they are managed accordingly. In this context, PM can be considered as focusing on action-oriented preemptive management, and, while encouraging improvement in visually aided management aspects, pursuing action-oriented approaches.

Approaches to an analysis of the relationship between product quality and the equipment follow.

3.3.2 Machine-Quality Analysis (M-Q Analysis Method)

Machine-quality analysis (hereafter referred to as M-Q analysis) activities are activities intended to analyze the relationship between machine precision (M) and quality (Q), and to obtain high quality and low costs. Instead of simply analyzing the interrelationships between machine precision and product quality, this analysis aims to analyze key factors, such as measuring instruments, manufacturing methods, and raw materials in relationship to equipment (see Figure 3.16).

When this type of analysis is carried out with cooperation among the production engineering staff, production shop floor supervisors, and maintenance engineers, it heightens awareness and understanding of quality among the maintenance technicians, and of maintenance technologies among the production workers. By means of collaboration among these groups, activities related to quality and equipment can be combined. As noted earlier, measures designed to totally eliminate quality problems are limited by human error and unrefined skills; most difficulties, however, result from mishandling of equipment, molds, and fixtures, inadequate skill levels of machine tuning and adjustment, and an inadequate daily maintenance routine designed to cir-

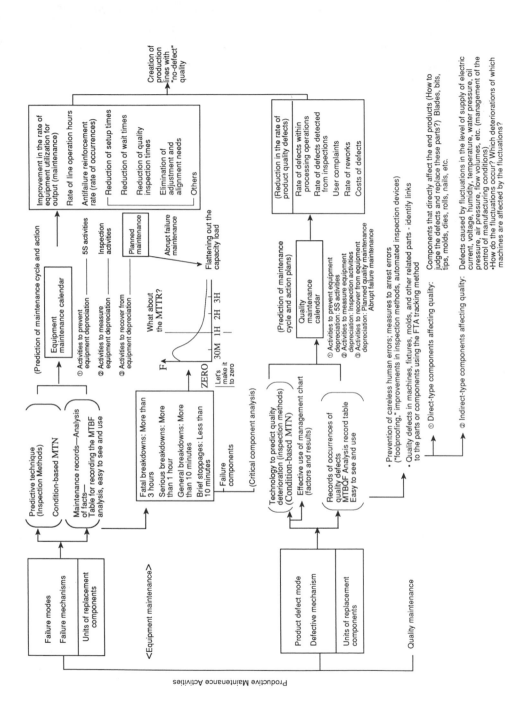

FIGURE 3.15 Concepts of Equipment and Quality Maintenance and Their Operational Basis

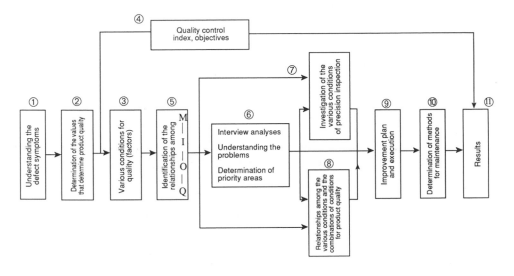

M — Machine, equipment
Q — Quality, product quality, intermediate product quality
I — Instruments, measurement instruments and equipment
Ma—Materials
O — Operation

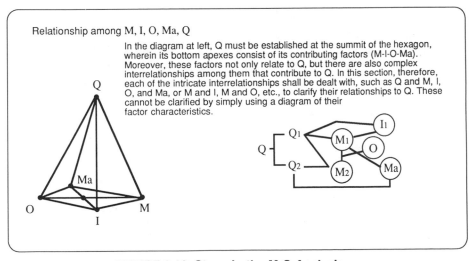

Relationship among M, I, O, Ma, Q

In the diagram at left, Q must be established at the summit of the hexagon, wherein its bottom apexes consist of its contributing factors (M-I-O-Ma). Moreover, these factors not only relate to Q, but there are also complex interrelationships among them that contribute to Q. In this section, therefore, each of the intricate interrelationships shall be dealt with, such as Q and M, I, O, and Ma, or M and I, M and O, etc., to clarify their relationships to Q. These cannot be clarified by simply using a diagram of their factor characteristics.

FIGURE 3.16 Steps in the M-Q Analysis

cumvent careless errors. Thus, it must never be forgotten that there are many issues to be dealt with in PM management.

3.3.3 Creation of a Foolproofing Concept Manual and Daily Routine Maintenance of Foolproofing Devices

Among the excellent techniques to highlight quality problems is the Japanese concept of "poka-yoke" ("foolproofing"). If this technique can be developed by shop floor personnel who are directly involved in potential problem areas, it could have a significant effect. "Foolproofing" refers to mechanical devices or setups at key areas in equipment or operation processes intended to build quality into the product while preventing careless mistakes caused by worker negligence. Its characteristics include mechanisms that (1) prevent a workpiece from attaching to a fixture when an operational error occurs; (2) prevent a machine from starting when there is a problem in the workpiece; (3) prevent a machine from processing when an operational error occurs; (4) automatically correct and restart processing when there is either an operational or processing error; (5) prevent quality failures by examining problems created in the preceding operational step by comparison with the subsequent operational step; and (6) prevent the succeeding operational step from starting when an operation has been overlooked or omitted in the previous step. Foolproofing techniques for this type of situation are (1) indicator signal method: using a lamp that signals problem occurrences, making them more noticeable by differentiating colors; (2) fixture method: a fixture that prevents a nonacceptable workpiece from fitting into position, and that prevents the equipment from starting if a nonacceptable workpiece is placed in position; (3) automation method: a method that automatically halts a machine when a problem occurs during processing.

Improvements made in foolproofing often appear simple when completed; they are not, however, so readily devised. If many of these improvements are expected to be derived from the practical wisdom of personnel on the shop floor, the workers must be thoroughly familiar with the relationship between the functions and quality of the end products on the one hand, and the functions and mechanisms of the equipment and fixtures on the other. Most of the useful ideas proposed in QC group activities are related to improvements in equipment, fixtures, and foolproofing for better quality assurance. When this occurs, it is a clear indication that shop floor personnel are becoming knowledgeable about these interrelationships. From its inception, "Total Participation PM" focused on this educational process, as well as on nurturing people who would become thoroughly familiar with equipment and fixtures, strengthening their ability to take the initiative in making improvements on their own. In many Japanese plants, hundreds or thousands of foolproofing measures and devices were developed for different products. But simply listing them as examples would not be sufficient. Rather, staff members need to create a "manual of the foolproofing concept" available for all to learn from and use. Past foolproofing logs should be classified by purpose and by shape of parts, so as to create a "how-to" handbook on the various methods available to deal with certain types of defects for any given part. An aggressive program should be launched to teach the concepts of foolproofing to as many people as possible on the shop floor. Moreover, there should be a map indicating where each of the foolproofing devices has been installed so that they can be included in a daily routine checkup to maintain them in perfect working order.

These are all quality maintenance activities that can be carried out on the shop floor (see Figure 3.17 for further information).

3.3.4 Zero Failure and Zero Quality Defect Activity through Use of an Equipment and Quality Map (Chart)

—CREATION OF A PERFECT PRODUCTION LINE—

The author once led PM activities in a vertically integrated petrochemical conglomerate that included a continuous process plant where raw materials produced in another part of the conglomerate were used to produce synthetic rubber that was ultimately to be used for the manufacture of automobile tires. In this plant, we created a map (chart) illustrating the relationship between equipment, breakdowns, and quality failure modes. This map was designed to analyze the interrelationships, establish maintenance management standards at the component unit level, and improve maintenance by identifying the various failure modes of each critical component: failure causes, presence of secondary failures, prediction of down time duration, spare parts availability, and emergency response methods.

We particularly sought to evaluate the effect of each critical component on equipment failures and on the quality problems of the processed products, as well as the reliability of the components themselves. Great progress was made by managing the maintenance program based on the overview derived from the results of these analyses. Figure 3.18 shows an example of this development.

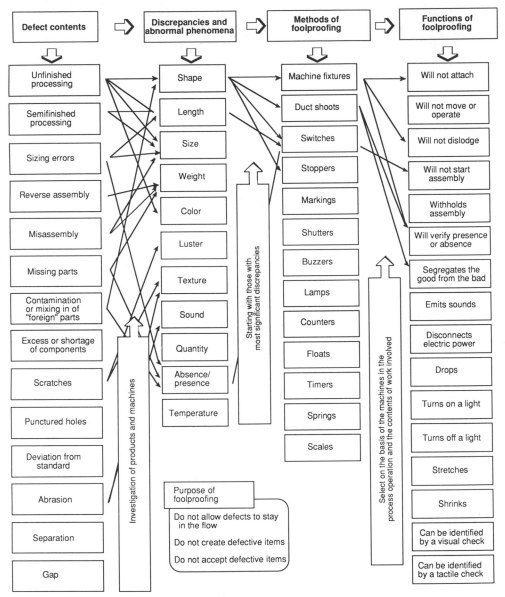

FIGURE 3.17 Specific Factors Related to Foolproofing Measures

3.3.5 Creating and Developing Quality Production Lines

In this section an example is discussed of how equipment-oriented quality maintenance activities were integrated into conventional quality control activities; how the weaknesses of quality control circle (QCC) activities were overcome; and how results were made even more effective. The firm discussed here has made utmost quality its company edict. It is a first-class business that once every five years has competed for, and several times has won, the Deming Award and the Japanese Quality Control Award, both with shining records. After winning these awards, and in response to new equipment sophistication and technological innovations, the company introduced PM with total employee participation. As one of its goals, moreover, this firm took on the challenge of competing for the "Best Business Environment with PM in Practice Award."

As part of its PM implementation strategy, under the slogan "TPM to Enhance Quality and Productivity," each of the company's plants developed a variety of unique and specific TPM activities. In particular, its personnel made strenuous efforts to establish quality maintenance activities for a 100 percent quality assurance program, i.e., building the quality production line, in which process the author played a leading role, assisting his colleagues together with the plants' personnel. Products made in these plants are comprised of many parts and components, with materials lists consisting of several thousand items, each contributing to the function and performance of the end products. To guarantee 100 percent quality for these items, it was necessary to identify the relationships between the quality characteristics of each individual item and the function and performance characteristics of the end products, and then to assure the quality of each.

- Problems arising from wear and deterioration of machinery components and from fixture wear would account for 50 percent of the total problems, but these can be avoided by preemptive measures.
- Human errors in assembly, handling, and setup account for 40 percent of the problems. By thorough training of production workers, these errors can be eliminated.

With these guidelines, we can deal with the task of creating the quality assurance production line (QL) using these quality maintenance development efforts.

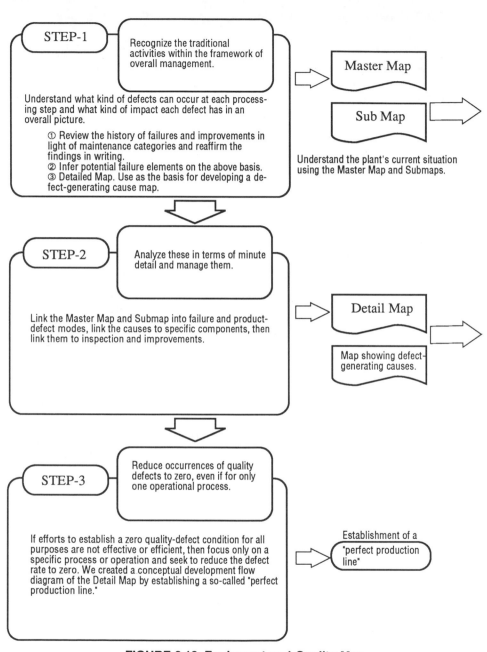

STEP-1

Recognize the traditional activities within the framework of overall management.

Understand what kind of defects can occur at each processing step and what kind of impact each defect has in an overall picture.

① Review the history of failures and improvements in light of maintenance categories and reaffirm the findings in writing.
② Infer potential failure elements on the above basis.
③ Detailed Map. Use as the basis for developing a defect-generating cause map.

Master Map

Sub Map

Understand the plant's current situation using the Master Map and Submaps.

STEP-2

Analyze these in terms of minute detail and manage them.

Link the Master Map and Submap into failure and product-defect modes, link the causes to specific components, then link them to inspection and improvements.

Detail Map

Map showing defect-generating causes.

STEP-3

Reduce occurrences of quality defects to zero, even if for only one operational process.

If efforts to establish a zero quality-defect condition for all purposes are not effective or efficient, then focus only on a specific process or operation and seek to reduce the defect rate to zero. We created a conceptual development flow diagram of the Detail Map by establishing a so-called "perfect production line."

Establishment of a "perfect production line"

FIGURE 3.18 Equipment and Quality Map

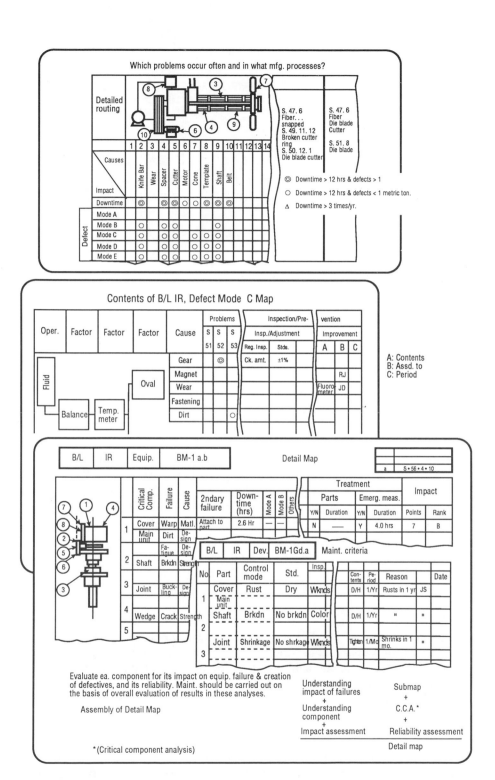

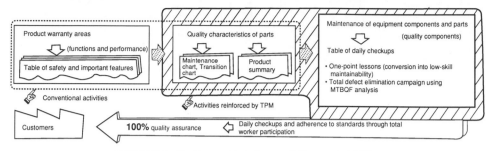

FIGURE 3.19 Items Reinforced Through TPM

(1) A Way of Thinking About Creating QL and Focal Points for Activities

(i) Creating QL. Through total participation, by performing "forward-looking management" (preemptive management), quality failures can be prevented. For this purpose we should consider:

Equipment-related causes: Identify in advance the factors and components that would impact on the quality of end products. Link their daily routine checkups and maintenance so as to prevent problems.

Human causes: Establish a one-point lesson for each of the operation elements that are prone to human errors and train the workers until they can work correctly. After equipment is installed, use as many foolproofing devices as practical so as to prevent human errors.

(ii) Items reinforced through TPM (see Figure 3.19).

(2) Dealing With the Issues

The various factors should be categorized based on analyses of past problems, know-how, and experience. Recognizing the characteristics of both the products and production processes, develop a program of production process maintenance and improvement activities, as shown in Figure 3.20, which presents a deployment chart of program development for creating QL.

(3) The M-Q Relation Analysis Table

A tabulated approach—the M-Q Relation Analysis Table—can be used to clarify the relationship between the required product quality char-

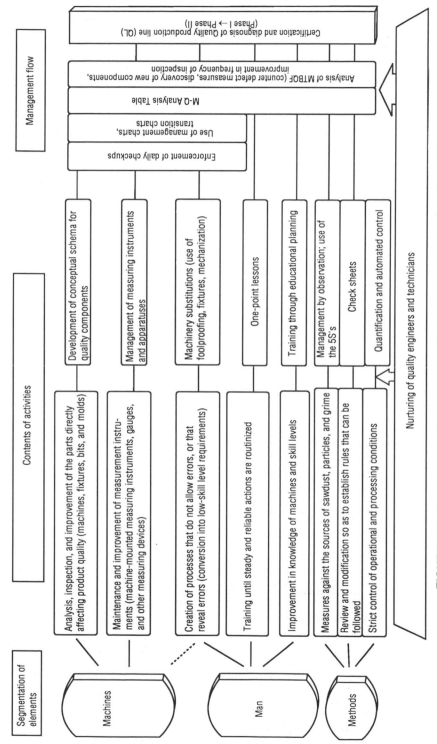

FIGURE 3.20 Deployment Chart for a Quality Assurance Production Line (QL)

99

acteristics and the production processes with regard to both quality enhancement and inspection. Since the tabulated approach enables assessment of the assurance level that can be delivered for each required quality characteristic, as well as relating assessment results to improvement activities, it makes possible evaluation of the specific production line's quality assurance level. An example of this table is shown in Figure 3.21.

(4) Management of Quality Components

Quality components are comprised of both direct and indirect types. A conceptual scheme of the quality component concept is given in Figure 3.22.

The direct quality component type refers to those equipment parts or components whose breakdown or problems would directly affect product quality. Material fractures, breakage, wear, and haphazard or last-minute maintenance of blades or bits are cases in point.

Since it is apparent that direct components could cause quality problems if kept in use beyond their maintenance limits, meters and measuring instruments are used to monitor intermittent changes and maintain standards.

There are three kinds of indirect-type quality components. The first is comprised of basic utilities, such as electricity, air pressure, and water. The supply and delivery of these utilities is usually assumed to be constant and stable. Certain elements, such as coolant water, can be readily checked for its rate of inflow or outflow. Careful observations, however, might show that deterioration of cooling capacity is occurring, caused by mineral deposits in the water. The surge pressure of air may also be decreasing, caused by too many interconnected air hoses. Coping with these problems requires asking consistent questions, such as: Are fluctuations or deviations occurring? Is there a consistent quality assurance program in place from the source of supply? What can be done to ensure that this program is effective?

The second kind of indirect quality component refers to those equipment problems that affect product quality, such as fluctuation in product precision due to the uneven revolutions of a drive train, caused by loosening or slippage of a V-belt. Other examples include distortion of a workpiece caused by misalignment of a revolving shaft, or a problem caused by a worn-out guide pin. These deteriorations occur over a long period of time and are difficult to detect or define in daily routine checkups. Thus, the degree to which these problems affect product quality greatly depends on the type of products and the manufacturing processes involved. Solving them by the application of a general formula is most difficult.

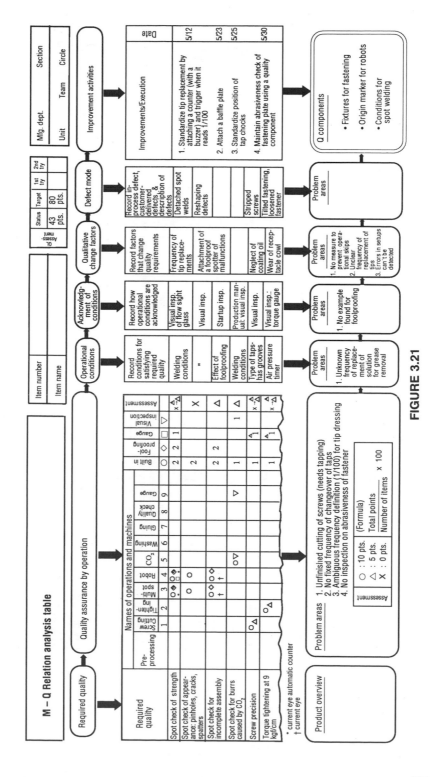

FIGURE 3.21

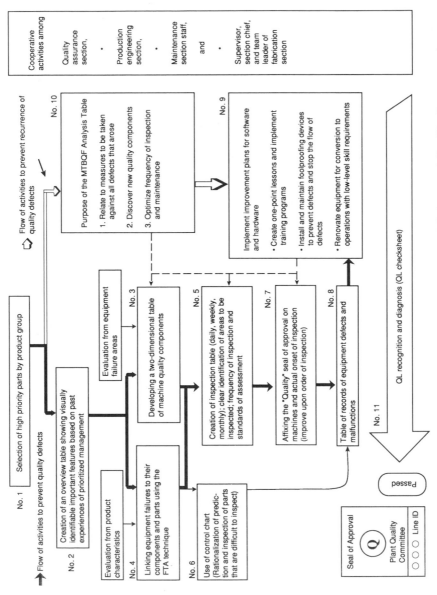

FIGURE 3.22 Eleven Steps in Developing the Conceptual Scheme of Quality Components

The third indirect quality component type refers to those numerous problems that are not usually considered to be related to quality degradation.

For example, a limit switch or a photoelectric sensor tube used for controlling material positioning may lose timing accuracy due to obstructing burrs or flakes from the previous workpiece. Or an electric relay may lose its timing accuracy due to burned-out contact points. Or one of many bolts may loosen, having a "snowballing" effect on other parts and components until the quality of the end product is affected. In these cases, moreover, discovery of the causal relationships is not simple, since the problems tend to affect many areas in a complicated chain reaction.

Thus, a thorough investigation and efforts at improvement are needed for every area that may affect quality, unless it can be determined that no causal relationship exists in a particular area. The key here is the degree of thoroughness. The problem with this approach, however, is that it is not always clear as to what the next maintenance step should be or how it should be carried out. Thus, even for this kind of component, specific track records should be noted for the identifiable components using past accumulated knowledge and the MTBF Analysis Table.

As noted before, these indirect components are characterized by difficulties in identifying the relationship between the component problem and the quality problem of the product. It is necessary to establish a ranking order of quality impact.

Figure 3.23 is an example of a quality component deployment table, following logical analysis steps for identifying quality components. This table was created by shop floor personnel who shared their experience and know-how. In the diagram, the relationship between equipment components and their quality characteristics are indicated by a circle ◯, and the number within each circle indicates the rank order of the strength of the relationship. The right-hand side of the table shows maintenance methods for the components.

As the condition of quality components changes, these changes cause product quality problems. A key step in maintenance activities is to determine which components should be maintained and how. With regard to the question of where the focus of maintenance activities should be placed, unless attention is paid to key pragmatic aspects, we may end up simply with a finer categorization of components or an elaboration of the details of maintenance standards. Aspects to be addressed include: (1) areas that lend themselves to preemptive action against quality failures through visual observation; (2) areas that allow for quality verification through checkups or other methods;

Class:	Case, blank, saw, polish, HT
Machine name:	SV equipment

Catalog no.	Saw cutter 005	Serial no.	
Manufactured:	April 30, 1990		Signature
Revised:	Month, day, year		Signature

Quality Component Deployment Table

Unit component	Component	Symptom	Saw-edge discrepancy	Saw alignment discrepancy	Oscillation of sawteeth groove	Pitch error	Surface coarseness	DPD size (large or small)	Manufacture Inspection Period	Manufacture Observation points	Manufacture Method	Replacement, minor repair	Maintenance Inspection Period	Maintenance Observation points	Maintenance Method	Repair Spare parts	Repair Method	Defect symptoms
Center	W nut	Looseness			○	③	○	○										
Center	Bearing	Wear		①	①	①	○	③	1/3M	Off-center alignment	0.005< dial	Replacement	1/6M	Bearing	Rotate by hand	Yes/present	Tighten	Fluctuation in precision
Clamp	Air pressure	Lowered	②	○	○	○	③	○	1/D	Pressure gauge reading	Eye	Adjust	1/6M	Swing	Measurement	Yes/present	Replacement	"
Clamp	Razor	Repair	①	②					1/M	Bearing	Align power conveyance gear	↑	1/Y	Clattering	Measurement	—	Adjust	Broken cutter or workpiece
Table	Limit "Dog"	Dislocation	③	③					1/W	Misaligned indicator on platform	Eye	Fasten					Adjust	Appearance of irregularities in the table advancement
Table	Groove congested by sawdust	Sawdust contamination							1/D	Foreign objects	Cleaning							Broken cutter or workpiece
Maintenance data	Control chart		○															Scratches between the tail stocks
Maintenance data	Transition chart																	
Maintenance data	Check sheet																	
	Remarks																	

Remarks:

○ denotes that strong relationship exists; the number indicates the rank in order of importance

FIGURE 3.23

104

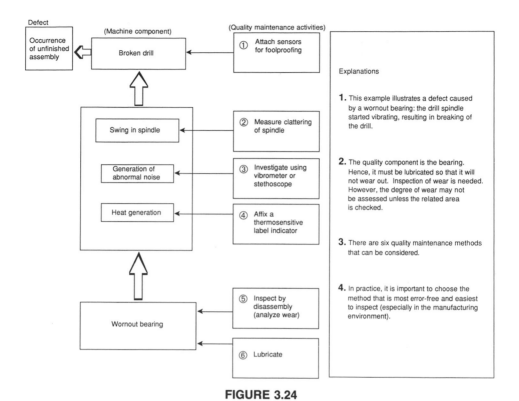

FIGURE 3.24

and (3) areas in which maintenance activities are easily carried out. Figure 3.24 gives an example of how this process can unfold.

3.4 ANALYSIS OF THE RELATIONSHIP BETWEEN COST (C) AND EQUIPMENT

The quality of equipment has a major impact upon the production cost of the end products. Yet the cost of maintenance must be lowered as much as possible, since there is a tendency to allow creeping increases in the ratio of the plant maintenance cost to the total production cost. In principle, however, a policy of simple reduction of maintenance cost is not necessarily the best one. If a total participation program of self-initiated maintenance can slow equipment deterioration, improved maintenance can extend the life span of equipment parts and components. By identifying signs of part and component deterioration and taking corrective measures well in advance, it should be possible to reduce breakdowns to one-fifth or one-tenth of their level prior to implementing the program. The absolute amount of maintenance costs, moreover, should be reduced as well. In many cases maintenance costs have been reduced by thirty percent while the number of unexpected breakdown occurrences was reduced to one-tenth of previous levels.

As the trend continues of replacing people with machines as the means of production, the effect of machinery on product cost has today become significantly greater. Are investments in new equipment, or in the expansion or the improvement of existing equipment, producing the kind of results that were initially expected? Did an error occur in the timing of the investment? Were there any errors in the design or construction of the equipment or in material selection or procurements? Were there any incidences of production startup problems after the plant was installed, causing delays of five to six months? Is the investment that was intended for manpower reduction achieving the results envisioned in the original plan?

These are crucially important questions for the management of production costs. In addition, analysis is needed on occurrences of losses and reduction in production levels caused by equipment breakdowns, as well as the related recovery and repair costs, the quality defects and various lot sizing-related losses, scrappage losses, and personnel expenditure-related losses, which are all caused by equipment problems. At the same time, it is important to understand the relationship between varying or changing equipment investment amounts and their maintenance and repair costs. Such comparisons would provide an important management checkpoint for assessing the effectiveness of maintenance activities.

Comparison would also allow us to evaluate the nature of the expenses: whether they were for production improvements, maintenance improvements, or preventive maintenance, or for the recovery of abrupt breakdowns. The percentage of the burden of equipment maintenance costs relative to the product cost has been rising annually.

People often say that equipment maintenance costs relative to production cost represent a mere three to four percent of the total, and hence that equipment maintenance costs can be left outside the scope of management focus. But when we listen more closely, it becomes clear that the production cost to which they are referring is the sum of the material cost and the processing cost. The cost of purchased raw materials is considered to be very difficult to manage as an accounting category in plant management, although reduction in the stock inventory of raw materials is very important. Nevertheless, there are many industries in which the subject of cost management for manufacturing enterprises focuses only on production costs. Moreover, when these same people speak about the cost of repairs, they tend to treat repair costs as the sum of external repair costs (which are accounts payable entry items) and materials purchased for maintenance, but they exclude the personnel costs of internal equipment-related departments. If the percentage figure based on these repair costs is seen as a dividend and the aforementioned product cost is seen as a divisor, they do not accurately show the real burden of the equipment maintenance cost.

Given the differences in plant type and production characteristics, the burden of the equipment maintenance cost is somewhere between twelve percent and forty-five percent of the costs incurred in a factory that are related to plant cost management, which primarily include production costs. (Even though the costs of raw materials and equipment depreciation expenses are generally included, many of these are administratively determined at the main office, and thus will not be discussed here.) The ratio of this equipment maintenance cost burden has increased in recent years. It should be noted that maintenance-oriented operations conducted by production workers are included in the equipment maintenance costs, based on the percentage of work that is considered to fall into this category.

When, as during a recession, profit-making becomes difficult, the first item targeted to reduce expenses—as any maintenance engineer knows—is usually the cost of repairs. This is so partly because of the former image of repair cost management as an undisciplined hodgepodge. Nevertheless, reducing repair costs does lower overall costs. Thus, maintenance engineers and adjustment and tuning technicians must be made aware of the magnitude of their roles in reducing production cost and should be motivated accordingly.

Attention should also be paid to the role of spare parts. The spare parts and materials stock inventory is maintained in the warehouse, where many spare parts compete for space. Often, however, urgent purchase orders must be placed to obtain items not found in the stockroom. Because of careless storage, the items are sometimes not ready for use (e.g., too rusty), requiring preparation time; or it may take twenty to thirty minutes to locate a needed item because of poor stockroom inventory management. These situations cancel the effectiveness of spare parts. Spare parts are on hand to reduce repair time. If a purchase order for a spare part can be placed after a breakdown occurs, then it does not need to be kept in stock as long as the time between its ordering and delivery is kept within tolerable limits. In any event, poor spare parts management often causes an increase in stock inventory costs.

The inventory rate of repair materials and spare parts will provide a yardstick for evaluating their management:

$$\frac{\text{Ending Inventory Amount}}{\text{Used Spare Parts Costs Per Month}} = \frac{\text{Inventory Rate}}{\text{(How Many Months in Stock)}}$$

If the inventory turnover rate is three to four months, the PM standard is quite high. A number of key questions should be asked with regard to analyzing the PM standard, so as to continue integrating and improving upon PM activities. These questions include: Are the regular stock items appropriate? Are the stock levels of the regularly stocked spare parts adequate? Are the procurement methods for spare parts standardized, such as the order point, fixed quantity, fixed frequency, or double bin methods? Are the spare parts maintenance procedures sufficiently simplified? Is the warehouse layout and its floor or space utilization efficient? Can a desired spare part be retrieved within a short period of time? Are the responsibilities of the inventory maintenance personnel clearly defined?

3.4.1 A Management System and Methods for Making Equipment Repair Costs Efficient

Repair costs that are organized and managed in a rational manner are indicators of the level of a plant's PM standards, and can be said to represent the system of maintenance itself. It is important to avoid a passive attitude that allows repair costs to rise while the equipment or the manufacturing processes are aging (resulting in frequent breakdowns). The key, rather, is a determined will, expressed in an awareness that, since X amount of money is being invested in equipment, the rate of failure and number of problems will be reduced by X amount. Concomitantly, it is expressed in the determination that since the equipment utilization rate will be increased by X amount, the energy cost savings shall equal it.

These things are obviously easier said than done. Nevertheless, an evaluation procedure for each of the priorities set for the manufacturing process or production line must be established, so that each expenditure for repairs and improvements is measured against manageable results in productivity. In essence, we should realize that an organized method for bookkeeping and repair costs management is a yardstick by which an entire system of productivity maintenance can be measured.

The repair costs should be systematically classified according to the source of repair and maintenance needs, whether for a specific object or type of item, so as to establish a classification of breakdowns. In ensuing analyses of actual failure occurrences, improvements should be made after conducting analyses of the equipment breakdown, the maintenance effect, and the maintenance cost analyses. These should be followed up by review of the maintenance plan, engineering and technical training, and reviews of the spare parts inventory, all of which should focus on efforts to plan maintenance so as to meet the challenge of overall cost minimization.

3.4.2 Creation of Standard Patterns of Maintenance Cost Using the MTBF Analysis Table

In the process of budget compilation, cost accounting departments traditionally make strong demands for the standardization of maintenance costs. These demands are not aimed at standardization for its own sake, but are motivated by the desire to spend maintenance funds effectively, to correct managerial errors manifest as operational or adjustment and tuning errors, or to correct ambiguities in planned maintenance periods. Given the general trend toward annual increases in the percentage of machine maintenance costs in plant maintenance costs, these demands can be readily understood.

Equipment that produces goods whose designs change frequently requires concomitant replacement or major modifications in three- or four-year periods. When such pieces of equipment are planned, their design and manufacture needs must be emphasized. Moreover, their overall life cycle costing and engineering evaluations must be reviewed within relatively short life cycles. These relatively short evaluation cycles need to be repeated for each successive planning period.

With regard to processing industries and industries that manufacture machinery and hardware intended for production and manufacturing applications, each investment in their various pieces of equipment tends to be enormous. Thus, their replacement is not simple. In such cases, equipment must be utilized to its maximum capacity, generating the most effective output with the minimum possible maintenance and operational investment. Since this equipment is used for 10 to 15 years, one seeks to identify the most economical and efficient patterns of maintenance expenditure utilization. The best approach for this evaluation and planning is based on maintenance records (MTBF

Analysis Table, see Figure 3.12), augmented by the maintenance cost transition table for machines or equipment, the tables of analysis and changing relationship between cost depreciation and maintenance costs, and the trend analysis of breakdowns by machines and equipment.

By examining the characteristics of plant equipment and machines, those pieces of equipment that show a high correlation with the maintenance cost should be chosen as the standard management measure. There is no need to select only one such yardstick; various criteria can be considered for each group of equipment. For example, the amount of electric power consumption, the consumption rate of other utilities, the actual operating days or hours, the frequency of model changeovers, and the volume of production are some of these management yardsticks.

This inquiry will be futile unless it directly affects actions to raise the standard of maintenance activities through the use of equipment-related costs. Moreover, unless there is a high maintenance management standard, there is not much point in even beginning to plan for establishment of standard patterns for maintenance costs (see Figure 3.25).

3.4.3 A Case of Repair Cost Reduction Activities

In a medium-size petrochemical processing plant, rising material costs and depressed product prices made it necessary to improve profitability. This in turn led to the need to review and improve repair cost management. Previously, reductions in the number of maintenance crew workers, increases in the number of machines and equipment, and an increased level of reliance on subcontracting (in an effort to cope with the overall aging of the plant facilities) had contributed to creeping increases in repair costs. In addition, periodic tuning, adjustment, and plant facility modification and enhancement work had tended to increase in scope, which together contributed to increased subcontractor costs.

A PM plan with total employee participation was aggressively introduced in this plant. As part of this effort, a five-year plan for reducing repair costs was also instituted. Annual increases in plant repair costs were estimated at ten percent or more, based on contributing factors arising from material costs, increased personnel costs, increases in the quantity of machinery and equipment, and measures taken for their depreciation. Using 1978 as the base year, a guideline of five percent was chosen to be the intended rate of annual increase. As improvement activities were enhanced, the amount of in-house

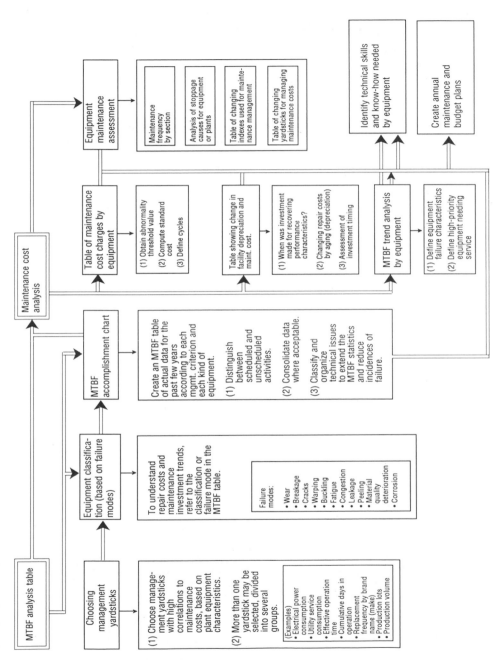

FIGURE 3.25 Schematic View of Maintenance Cost Based on the MTBF Analysis Table

111

manufacturing was steadily regained. At the same time, ongoing education and training enhanced the maintenance skills of the production workers. By these efforts, a result significantly lower than the five percent guideline was achieved.

Table 3.2 shows the key points and a checklist for reducing repair costs.

TABLE 3.2 Checklist of Key Points for Repair Cost Reduction

I. **Discontinue**

1. Stop (or postpone) repairs

  1-1) No impact will be felt on safety, quality, or production
  1-2) No problems will arise (rely on confidence)
  1-3) Corrective maintenance will suffice
  1-4) The machines are stopped (not in use)
  1-5) Abandon repair efforts if stoppage occurs

2. Convert periodic checkups into postfailure reactive repairs

  2-1) Discover repair methods to prevent reduction in production quantity
  2-2) Discover that the conversion has not caused reduction in production quantity

3. Extend length of maintenance periods

  3-1) Reinforce daily routine checkups
  3-2) Alter manufacturing conditions
  3-3) Alter PM standards based on accomplishments
  3-4) Alter PM standards in absence of data on accomplishments (rely on confidence)

II. **Reduce (Economize)**

1. Attack the failure source

  1-1) Reinforce
  1-2) Alter the system
  1-3) Alter the designs
  1-4) Alter the materials
  1-5) Strengthen components
  1-6) Alter components
  1-7) Pay careful attention to operational details
  1-8) Alter positions (layout)

2. Alter and curtail repair methods and scope

  2-1) Repair selectively

  2-2) Limit scope of repairs to local areas
  2-3) Alter repair methods
  2-4) Optimize quality, materials, and precision of repairs

3. Convert to and emphasize in-house manufacture

  3-1) Concentrate and boost in-house manufacturing capabilities
  3-2) Switch to in-house overhauling of high-revolution machinery
  3-3) Switch to in-house overhauling of measuring instruments
  3-4) Switch to in-house repairs for others

4. Reduce spare parts and purchase of maintenance materials

  4-1) Recycle and reconstruct parts and materials
  4-2) Share the use of parts and tools
  4-3) Use idle machines as sources of parts and components
  4-4) Review existing stock inventory standards (reduce inventory)

5. Change vendors and procurement methods

  5-1) Change vendors
  5-2) Review delivered goods
  5-3) Change from patronized vendors to bid winners
  5-4) Negotiate

6. Reduce steps in the manufacturing process

  6-1) Utilize tools and mechanize
  6-2) Review process layouts and setups
  6-3) Reduce idle workforce
  6-4) Review manufacturing methods, raise precision levels of production plans, and integrate similar processes

III. **Summarize (Integrate)**

1. Transfer responsibilities and convert to cooperative shared responsibilities
2. Check if budgeted item entries and category identifications are correct
3. Check if inspection time is appropriate

TABLE 3.3 Maintenance Impact on Firm Economy

MAINTENANCE IMPACT ON FIRM ECONOMY				
	Maintenance cost (million dollars)	Percent of sales	Percent of net income	Percent of fixed assets
				at cost	after deprec.
Du Pont	$310.9	5.9%	53.1%	5.0%	12.9%
Union Carbide	244.9	6.2	84.2	4.9	12.3
Dow Chemical	151.0	4.9	54.8	4.6	8.5
Diamond Shamrock	31.3	4.8	62.0	3.7	6.8
U.S. Steel	771.2	11.1	236.7	7.4	18.3
General Electric	319.5	2.8	54.6	6.5	13.5
Gulf Oil	260.0	3.1	32.5	2.3	4.8
Goodyear Tire & Rubber	209.3	4.5	113.3	6.8	13.3
Eastman Kodak	160.4	4.0	24.6	4.8	9.4
International Paper	111.5	4.8	69.8	5.0	9.5
Johnson & Johnson	44.3	2.8	30.2	7.4	12.3
Beatrice Foods	39.3	1.4	44.0	6.0	9.9

(From a paper by G. Peter Longo of the Italian IFAP, Third European Conference on Maintenance, Stockholm, Sweden, 1976.)

An excerpt from a report on the state of maintenance costs among well-known international firms, "Maintenance Impact on Firm Economy," from the 1976 proceedings of the Third European Conference on Maintenance, is found in Table 3.3.

3.5 ANALYSIS OF THE RELATIONSHIP BETWEEN DELIVERY (D: DELIVERY DATE) AND EQUIPMENT

Delivery is the most direct factor that links a factory with its customers, regardless of whether the factory is involved in made-to-order goods or mass production. In parts and assembly manufacturing plants, management problems occur most commonly with regard to delivery-date management. In an environment where Just-In-Time is practiced and where the work-in-process inventory is kept to an absolute minimum—as exemplified in the Toyota Motors' production method—problems in delivery dates cannot be eliminated unless the standards of plant management are improved. To accomplish this, production equipment must be continuously maintained in such a way that on-going production requirements can be met at all times.

Each manufacturing step delivers only what is needed, when it is needed, and only the exact quantity needed. Each production step must meet these requirements, making it possible for each subsequent process to do the same. Should frequent failures occur, or should inferior quality outputs be produced by the manufacturing equipment, it is not possible to deliver goods on time.

If the causes of missed delivery dates or frequent urgent, special deliveries are investigated, it appears that most are caused by equipment or quality problems. Aside from missed delivery dates that result from poor management, difficulties in due date schedule management are intrinsic to some industries. These may be caused or affected by high demands concentrated in particular time periods, acceptance of orders above and beyond available plant capacities, or concentration of deliveries in particular time periods. In many cases the confusion in plant activities resulting from these intrinsic difficulties is equated with poor marketing, but this is an inaccurate perception that perpetuates a vicious cycle.

Missed deliveries caused by frequent production of defective products or by equipment breakdowns may become inadvertently sanctioned by efforts to absorb lost time through overtime work. But repeated overtime work causes cumulative worker fatigue, resulting in increased absenteeism, whereupon plant administrative supervisors may be seen struggling to manage the job assignments. To end this type of vicious cycle, it is necessary to institute thorough daily equipment maintenance, including meticulous tuning and minor adjustments. It is important that this daily routine focuses on reliable preventive maintenance in the key areas where sound equipment conditions have a direct impact on product quality, as well as on the prevention of abrupt breakdowns. The higher the volume of orders received, the more assiduous the efforts must be to manage daily equipment maintenance.

In some industries, people may opt to use buffer inventory of end-products or semifinished goods on the basis of speculative production increases. But this is only an easy way to inflate losses. For a realistic analysis of the delivery situation, the following should be carefully investigated: (1) missed deliveries caused by the instability of production schedules attributable to equipment problems; (2) the occurrence of overtime situations despite no incidences of missed deliveries; and (3) the amount of work planned. The impact of equipment upon due date management must be understood, in view of the fact that worker absenteeism and defective products are often due to equipment problems.

3.6 ANALYSIS OF THE RELATIONSHIP BETWEEN EQUIPMENT AND SAFETY, ENVIRONMENT, AND POLLUTION (SEP) FACTORS

Safety, environmental protection, and pollution control are of cardinal importance in plant management. Catastrophes such as fires or explosions resulting in environmental destruction can wreak havoc on the surrounding region—and ruin the company as well. PM activities in a large-scale processing industry should focus on management of these critical areas.

In the mechanical assembly type of industries, new machines may be serially introduced as production lines are laid out and installed. In contrast to processing industries, numerous ongoing changes, replacements, and modifications are made in the plant's equipment. As a result, accidents can happen because workers are unfamiliar with technical operations; in fact, these circumstances make it highly likely that this type of accident can occur.

Since there are well-established academic disciplines systematically focusing on studies of safety and environmental pollution, these areas will not be discussed here. Nevertheless, because machines comprise the major means of production today, and because problems in equipment are directly related to both safety and accidents, plant security evaluation and equipment management methods should be considered.

3.6.1 A System Concept for Plant Security and a Systematic Method for Its Evaluation

Conventional approaches to equipment security have been strongly influenced by a heavy dependency upon measurement-based mechanisms such as an interlock system. Such systems may bring about unduly protracted operations or may allow the entire equipment group to depend upon the interlock system's reliability. This approach, therefore, cannot dramatically improve the plant's reliability as a whole.

Such being the case, we have briefly discussed a holistic view of equipment security perceived as a system that emphasizes evaluation of its importance and reliability. This method — called plant security evaluation (PSE)—requires that ways be found to solve problems; in this way, the reliability of the system as a whole can be improved through PSE (see Figure 3.26).

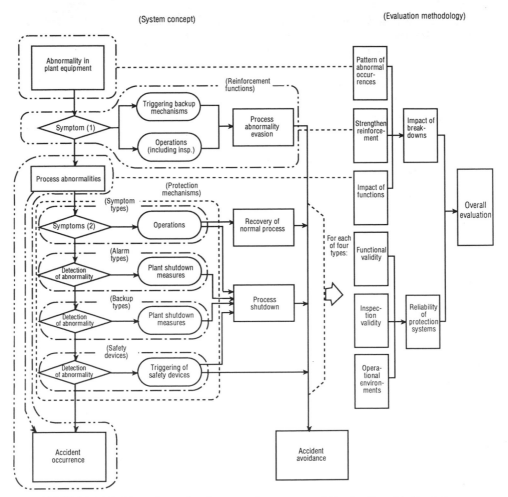

FIGURE 3.26 Plant Security System Concept and Evaluation Methodology

1. Instead of regarding security systems as mere measurement device mechanisms, the numerous methods for accident prevention (such as operational maneuvers, backup machines and equipment, measurement devices, safety devices, etc.) should be considered from an integrated perspective.
2. Security systems should be considered in a total developmental context from the onset of an abnormal condition to the occurrence of a disaster. Such a conceptualization would make possible quantitative evaluation of the scenario from a variety of viewpoints.
3. Proposals for improvement should be reflected in the construction

and maintenance of subsequent plant installations or in other similar plants.

4. The evaluation process and results should be utilized in both the plant operation and practical training of the maintenance crew.
5. Those problems whose solutions require more sophisticated technology should serve as the subjects for investigation into equipment diagnostic technology on a companywide basis. Furthermore, those problems requiring advanced research in order to understand the reasons for breakdowns should be dealt with in a similar manner.

3.6.2 Procedural Steps for Implementing the Operational Specifics of the Plant Security Evaluation (PSE) System

1. CHOOSING SUBJECTS FOR INVESTIGATION

With regard to the extent of potential disasters, security systems within a plant vary greatly as to importance and reliability. Using information derived from daily operations, maintenance activities, data from other systems, and instances of accidents in other plants, unstable equipment should be the first selected for investigation.

The selection procedure begins by subdividing a plant into sections, and identifying a specific subsystem or subsystems for each section. The next step is to define the goals for each subsystem and determine the subjects of investigation.

2. METHODS FOR RELIABILITY EVALUATION

(i) Analysis of causes of accidents. Thoroughly research the causes of accidents selected for further investigation, and organize the data by incorporating them into a table of causal relationships of abnormalities.

(ii) Evaluation of the impact of breakdowns. Each cause of breakdown should be reviewed and assessed (according to the three criteria that follow), and the data organized into a table showing the extent to which breakdowns affect each individual case, and into a worksheet evaluating the extent of the impact of the breakdowns.

1. Patterns of abnormal occurrences: frequency of machinery and equipment breakdowns, predictability, and effectiveness of inspection.

2. Strength of reinforcements: presence or absence of spare machines and equipment and their level of use during operation, aimed at avoiding equipment failures and processing abnormalities.
3. Impact of assigned function: the time it takes for equipment malfunctions or process abnormalities to develop into an accident, the magnitude of danger once an accident occurs, and the degree of difficulty of recovery.

After assigning a point score rating to the criteria, focus on those areas that scored higher than a standard level (a warning that they need attention), clearly identify their problems, and use them as the subjects in the next evaluation cycle of the security system's reliability.

(iii) Evaluation of the security system's reliability. Security systems can be thought of in terms of the following:

1. Symptom-oriented systems: Based on analysis of given symptoms, the system either recovers its normal state or shuts down the equipment where such functions are part of the equipment's operation.
2. Attention-getting systems: When it is not possible to implement a symptom-oriented system, this type detects abnormal conditions and terminates the operation of the equipment. This is a primary measure.
3. Backup-oriented systems: When attention-getting type systems do not function well, this type detects abnormal conditions and terminates the equipment operation. It is a secondary measure.
4. Safety devices: When even a backup type system fails to respond, this type protects the equipment from accidents. This type includes such devices as the safety valve, rupture disk, etc.

Evaluation should be carried out based on the following three criteria, and the data gathered should be organized on a security system reliability evaluation chart.

1. Effectiveness of functions: reliability of operation and effectiveness of safety devices.
2. Effectiveness of inspections: timing and thoroughness of inspections.
3. Environmental conditions: heat, vibration, corrosiveness, etc.

(iv) Overall evaluation. Organize evaluation results derived from steps (ii) and (iii) into lists of overall evaluation and countermeasure tasks.

Subsequently, taking into consideration the balance among the items in each of the lists, evaluate the system as a whole. By analyzing the overall summary table obtained in this way, it is possible to understand clearly the effective points for improvements. The actual tasks to be undertaken for improvement should be initiated in those areas with the highest level of impact from breakdowns or those with the lowest level of reliability in the protective systems.

These steps should be discussed in the PSE investigative meeting. Due to the diversity of the problems, the meeting should include personnel with a variety of experience and expertise.

3.6.3 Relationship Between Overall Equipment Management System and PSE

In our case example, the goal was to attain the level of perfect production through total employee participation. In pursuit of this goal, the operations, maintenance, and engineering departments jointly created the various kinds of equipment maintenance systems, as shown in Figure 3.27. It is worth noting here that the security evaluation system used is comprised of an assortment of systems that make up the equipment maintenance systems, which have together been achieving good results in conjunction with other equipment management systems.

In mechanical and assembly industries, a significant number of accidents are caused by human error. A number of accidents result from cavalier attitudes toward the need for training on production equipment (despite improvements made upon them), and others result from sheer lack of knowledge about the function of the equipment.

Other accidents may be caused by malfunctions of interlocking mechanisms for machine enclosures, covers, or interlocking mechanisms within the machines themselves, which in turn are caused by vibrations, loosened parts or components, and accumulation of dust and grime. For interlocking mechanisms, phototubes, limiter switches, pressure-sensitive switches, timers, indicator lamps, buzzers, solenoid valves, torque limiters, and other kinds of devices are used. Too often, however, these devices are not immune to breakdowns. Component problems in these devices at critical function times lead to accidents. PM must deal with these malfunctions via a preplanned approach.

Even for processing industry production lines with a heavy concentration of equipment, there are many cases of line stoppages caused by problems of the previously-mentioned components, and requiring

Safety Management Information System

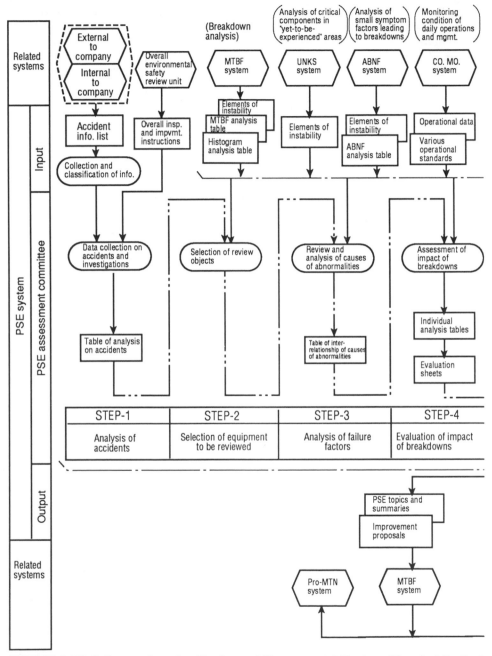

FIGURE 3.27 A Comprehensive Equipment Management System Aimed at Perfect Production

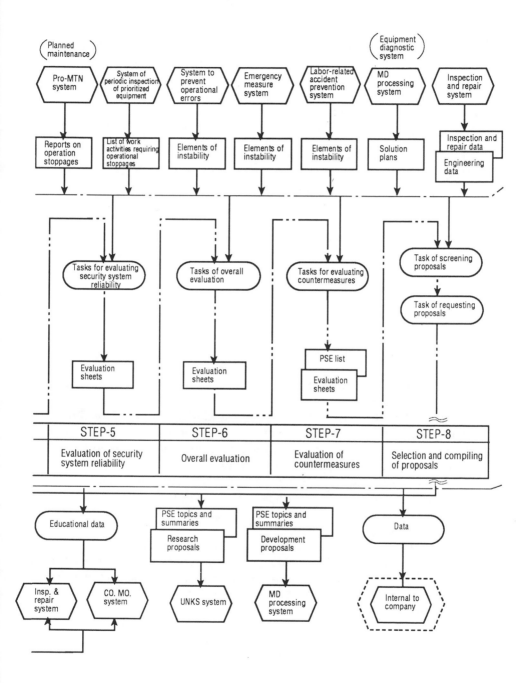

STEP-5 | STEP-6 | STEP-7 | STEP-8
Evaluation of security system reliability | Overall evaluation | Evaluation of countermeasures | Selection and compiling of proposals

special attention. Instances have occurred, for example, in which an error in positioning a phototube, or carelessness in installing one, caused a malfunction of the interlocking device, which ultimately led to a major accident. Another type of incident was related to equipment protection devices such as the torque limiter. Because of rigorous production schedules and their hard use beyond the design limits—ignored amid demands for production—the torque limiters would break often enough to disturb the work schedule, causing the workers to tamper with the limiters just to halt the nuisance of stoppages. The resultant overload led to a major accident. Safety inspection conducted from the PM engineer's perspective therefore must be carried out with a specialist's knowledge of these devices.

3.7 RELATIONSHIP BETWEEN MORALE (M: MOTIVATION) AND EQUIPMENT

3.7.1 The Idea of the 5S's

The question as to whether workers have an interest in and attachment to their machines and are using them with care can be answered by analyzing the realities of the management of the 5S's in the plant. These are: *seiri:* order; *seiton:* tidiness; *seiso:* cleaning; *seiketsu:* personal cleanliness; *shitsuke:* discipline. Notwithstanding the scale and scope of plant operations, or regardless of how nicely a supervisory chalkboard may be set up in a mechanical fabrication and assembly plant, if the operation console panels are dirty, or if the shop floor is messy and machines are smeared with grease and grime, with no sign of recent maintenance, then quality production cannot be sustained over time, and the equipment will not last. No matter how sophisticated the machines and systems may be, people operate them, and a qualitative change in human behavior must occur to improve productivity through PM.

Throughout Japan notice boards extol the virtues of the 5S's, including order, tidiness, and cleaning. But in reality, there are vast discrepancies between the most and least successful of these efforts.

The management of the 5S's seems simple on the surface, but its practical implementation is difficult. In a plant where people are capable of rigorously managing the 5S's, regardless of what management methods may be introduced, these personnel know how to take advantage of them. This is because 5S management is not something that can be accomplished through the contributions of a few extraordinarily talented people. Rather, it can be successful only through individually motivated initiatives that apply to all employees with regard to order, tidiness, and cleaning. It should be clearly distinguished from a type of hypothetical management that merely promotes increased use of an equipment cleaning crew. This is meaningless in our understanding of 5S management. The 5S's are in essence the foundation of all plant activities.

A revised plan to be proposed here for implementing the 5S's consists of two aspects or types of plans. The first resolves problems once the revision plan is adopted. For example, if a limiter switch of a certain machine breaks down repeatedly, replacing the switch with one that was improved after investigation of the problem can eliminate future switch breakdowns. For this type of revision, talented engineers can solve the problems as they occur.

The other plan aims at heightening the efficiency level of a certain machine, but unless certain minimum work, inspection, or spare parts management standards are created through the research done by line and staff people alike and are adhered to by everyone in the plant, no discernible results may be seen from efforts made to accompany the revision proposals. In any plant, there is an enormous number of this type of revision plans. As noted before, they are significantly affected by individual self-initiative and motivation. Management of the 5S's is but a typical example of this type. 5S management consists of tasks that can be performed by anyone who is motivated to do so; therefore, the quality level of a plant can be determined by first providing a set of 5S management standards to a plant, which is the very first step toward implementing PM, and letting people practice them. If the rules of 5S management are not adhered to, then despite all subsequent efforts to introduce management standards, they may be of little use in that environment.

It should also be noted that the level of this management quality determines the quality of products manufactured in the plant. In other words, through this 5S management people learn the foundations of maintenance. Thus, 5S is often called the barometer of plant maintenance, and constitutes the foundation for sound plant activities. What really matters on the manufacturing shop floor—even more than the parts and products produced—is the level of shop floor maintenance quality. In the final analysis, shop floor activities are crucial to management of the manufacturing enterprise.

3.7.2 Concepts of 5S Activities

The concepts that comprise 5S activities tend to be overly didactic. This is because the activities are not centered on results, but rather they emphasize people's behavioral patterns, such as the elimination of unnecessary items from the work environment or the cleaning and neatening of equipment. Consequently, the activities are of a kind that make quantitative assessment of their effectiveness difficult. When 5S activities are promoted, questions are often raised such as "Which is more important, 5S activities or work?" or "How much earnings would 5S activities bring?"

5S activities are intended to qualitatively change the ways in which people think and behave, and, through these changes, to alter the quality of both equipment maintenance and the work environment. Before 5S activities are begun, everyone involved should clearly understand and codify in writing the goals and meaning of the 5S's for each individual company and work environment. Furthermore, since 5S activities must be carried out with determination and concerted effort, it is helpful to devise slogans. Unless the specific rules are made highly pragmatic, it is hard to put them into practice.

Examples of Slogans

- Never make a mess
- Never spill
- Never allow untidiness
- Clean immediately when anything becomes dirty
- Rewrite when writings become illegible
- When anything peels off, stick it back on
- Time must be made available for the 5S's
- Abide by established rules
- Execute immediately

The 5S's are comprised of result-oriented goals, the definitions of the 5S's themselves, and the 5S activities that unite these elements. The following explains the order of the relationships among this triad.

GOALS OF THE 5S'S

Safety and the 5S's

Whenever people gather in a work environment, regardless of how small the size of the group may be, their concern for safety prompts them to write relevant slogans on display boards, stressing at least two key words: order and tidiness. These placards serve as an important reminder of key safety activities, such as wearing hard hats or safety shoes, taking precautionary measures when transporting materials, keeping passageways clear, and other similar mundane acts. Other specific 5S activities relate to oversight linked to safety and accident prevention, such as plant fires caused by oil leaks, worker injuries resulting from slippery shop floors, or environmental pollution caused by dust or oil fumes. In a work environment where 5S activities are promoted, a reduction can often be seen in the number of worker injury cases, even though personnel may not be verbalizing safety or caution slogans.

Efficiency and the 5S's

First-rate chefs and carpenters take good care of their tools. Rust or chipped edges on their cutting tools are unthinkable; these tools are kept sharp and clean. The kind of work exemplified by first-rate artisans and technicians is performed with efficiency and quality. Not only do they take good care of their tools, which are arranged for maximum accessibility and with minimum exposure to damage, they try to improve them for ease of use as well. To carry out the 5S methods, the segmented management technique based upon time-differentiated classifications of equipment should be applied. For those items whose management attention intervals are specified in minutes or tens of minutes, the contents of 5S management should be rotated in such a way that everyone will be ready to carry out their assigned 5S duties at once when the "Start!" command is given.

5S in 3 minutes	Once a day
5S in 10 minutes	Once a week
5S in 30 minutes	Once at the end of a month or at a specified time
5S in 2 hours	On a specified date or at the end of a year

For example, the time of "3 minutes" in the above category "5S in 3 minutes" denotes the kind of 5S practices that can be assigned to and clearly defined for everyone, while the amount of time is sufficient to accomplish something significant once everyone is trained. Using this method, 5S practices may become highly efficient.

Quality and the 5S's

Grime, dust, or hair often contribute to product defects of electronic equipment and precision machine components or parts; in addition, sawdust or burrs will degrade fabrication precision levels. In an assembly manufacturing environment, unrelated parts can fail and interfere with assembly, wrong parts or components can be used, and mismatched deliveries can occur. These problems often result from a lack of cleanliness, tidiness, and discipline in keeping things in designated places. In the close relationship between the 5S's and quality, the 5S's acts as quality's linchpin.

Breakdowns and the 5S's

Oil pressure reservoir tanks filled with solid deposits or air equipment with filled drainage pans often cause the very peculiar problems called the "Monday maladies." Other problems caused by delays in implementing 5S practices include the clogging of lubrication fluid pipes, metal burns caused by broken parts, burned motors caused by dirt-clogged cooling fans, malfunctioning of limiter switches caused by sawdust-clogged mechanisms, and accidents caused by malfunctions of dirt-smudged phototubes or by errors in selecting correct switch buttons because they were unlabeled.

Obviously, if 5S practices are not in place, nothing of importance works well, be it safety and accident prevention, efficiency, quality, or equipment failure prevention.

Promotion of Morale

As in a work environment where high worker motivation goes hand in hand with the attitude that established rules will be strictly adhered to, so, too, the practice of the 5S's is directly linked to promotion of morale. This means that each individual worker takes on assigned responsibilities and creates an environment where everyone can participate with dedication and diligence.

Cost Reduction

As mentioned before, measures taken against oil or air leakage problems, or measures taken to reduce the level of inventory, can contribute greatly to cost reduction.

In the final analysis, the 5S's are those specific activities that form the foundation of any manufacturing enterprise.

TABLE 3.4 Meaning of the 5S's

Meaning of the 5S's (5 words starting with "S")	Definition	Examples of results	Objectives
Order (Seiri)	Distinguish the necessary from the unnecessary and remove the latter	• Reduce on-hand inventory • Use space efficiently • Reduce incidents of lost or missing items • Stop leakage of oil, air, etc.	Reduce costs
Tidiness (Seiton)	Determine layout and placement so that any item can be immediately found when needed	• Eliminate losses caused by searching for needed items • Eliminate unstable conditions	Improve efficiency
Cleaning (Seiso)	Eliminate litter, dirt, and foreign matter; keep environment clean	• Maintain and improve equipment functions • Clean and inspect key equipment areas	Improve product quality
Personal cleanliness (Seiketsu)	Keep the environment clean to maintain health and prevent pollution	• Improve the work environment • Eliminate causes of accidents	Reduce number of breakdowns
Discipline (Shitsuke)	Train people to implement decisions	• Reduce incidents of carelessness • Abide by rules • Foster better human relationships	Assure safety and pollution prevention
			Raise morale

3.7.3 Meaning of the 5S's (see Table 3.4)

ORDER (SEIRI)

Order means managing the flow of materials to avoid stagnation. The difficulty lies in distinguishing between necessary and unnecessary items, since there is a tendency to think that at some time in the future each item could become necessary. The distinction, therefore, should be based on the degree of necessity at the very beginning.

> *Rarely used items:*
> • Items that are not used even once per year
> • Items that are used once in 6 months to 1 year
> *Occasionally used items:*
> • Items that are used once in 2 to 6 months
> • Items that are used at least once a month
> *Often used items:*
> • Items that are used at least once a week
> • Items that are used at least once daily
> • Items that are used at least once per hour

If the items are differentiated according to this type of classification, it is possible to decide whether or not to discard those items that are not used once per year or even less frequently. This classification method also allows restriction of the amount of inventory, as well as enabling determination of the number of items that are necessary to be kept on hand at appropriate sites. Tens of metric tons of items could be discarded through such a campaign. A definitive program for cleaning dirt and grime can also be implemented for these items, and measures can also be taken to combat leakage sources of various types.

There are various kinds of leakages, such as air, water, electricity, oil, sands, shot beads, steam, raw materials, or even finished products. But the losses attributable to leakages are significant. For instance, just the corrective measures against air leakages from air compressors may result in the idle capacity of one air compressor unit, or we know of examples in which a repair for an oil leakage problem resulted in halving the amount of oil needed for the machine involved.

Periodic campaigns in which everyone participates are also necessary, as well as continued engineering improvements.

TIDINESS (SEITON)

Tidiness refers to determining the layout of items and how many pieces should be placed in appropriate locations; this is done once; only the necessary items remain as the result of the *seiri* campaign.

Those that are not used → Discard
Those that are seldom used → Place them in distant areas
Those that are used occasionally → Store in designated areas
Those that are used often → Place them within the work area or on workers' tool belts

---

These methods should work well. Another important question to be considered is how soon needed items can be brought from storage; the answer also determines the number of units to be kept on hand.

Those items that can be retrieved quickly → Keep only the minimum necessary quantity
Those items that take longer to be retrieved → Keep some extras

To achieve tidiness in a plant, it is important to clearly define passageways and work area demarcation lines. As a rule, physical objects should be placed in straight, perpendicular, or parallel lines, but placement should be preceded by the design of a layout based on considerations of the efficiency of work to be done.

The placement of physical objects requires ingenuity and keen observation. How successfully this is accomplished determines whether or not there is a sense of purpose in the work environment. Storage shelves and bins should be built and arranged in such a way that clearly marked stored items are both accessible and easily returned.

Establishing tidiness in the work environment depends upon a study of work efficiency. The amount of time needed to retrieve an item and return it to storage should be measured and the results can then be used as a yardstick to evaluate the level of tidiness (e.g., a 30–second retrieval for documents and tools).

CLEANING (SEISO)

The women volleyball players of Tokyo Olympic Games fame were called by journalists the "witches of the Orient." During practice sessions prior to the games, each of these athletes kept a handkerchief at the ready around her waist, and whenever sweat dripped onto the floor, she wiped the spot, to avoid slipping on a sweat-covered surface.

In the army, soldiers keep their weapons cleaned and well polished, ready for use whenever needed. Similarly, after a strict cleaning policy was implemented at one plant, the problems related to mechanical control systems disappeared.

Cleaning means more than simply cleaning up the general environment. It also means guarding from dirt every single part of each item used, upon which our livelihood depends. In other words, cleaning is a form of inspection. The problems caused by grime, dust, and foreign matter are related to all aspects of accidents, defects, and malfunctions. The recent tendency among young people to disdain dirty work environments is affecting personnel recruitment.

Cleaning should not be carried out by hired contractors. Rather, it should be an undertaking of the plant's employees, with responsibilities assigned to everyone. For large and public areas, a method of fair, rotational assignments is recommended.

"Color conditioning," i.e., choosing a pleasant paint scheme for the equipment and work environment, is an important element in creating a light and neat atmosphere. In the past, dark color schemes were chosen to hide dirt, but lighter colors have recently begun to be used. Sometimes there is a tendency, however, to use paint where inappropriate, so as to conceal stains and grime. The only positive aspect to this practice is that it at least constitutes the first step toward promoting workers' care and concern for their work environment. From the PM perspective, we prefer polishing machines and equipment rather than painting them from the outset.

PERSONAL CLEANLINESS (SEIKETSU)

Personal cleanliness often refers to hygienic conditions for food and drink, such as the facilities for hot water for tea. Recently, an increasing number of shop floor workers have begun to wear light-colored smocks. Since grime and dirt are readily seen on these smocks, the level of personal cleanliness can be easily judged. These cases exemplify the goal of making problems of personal cleanliness readily apparent.

In considering the environment and pollution in terms of the people affected, investigation of this subject expands into several other areas, all of which should be understood from the 5S perspective. These include dealing with oil mist, dust, noise, paint-thinner-type oil products, and poisons. Some enterprises promote placement of flowers in the shop environment. Thus, the idea of personal cleanliness embraces the upkeep of a generally clean and pleasant atmosphere.

DISCIPLINE (SHITSUKE)

In the shop floor environment many problems arise from what workers wear and how they care for themselves, such as injuries caused by neglecting the use of protective headgear or safety shoes, or accidents caused by workers' hair or gloves becoming tangled in machines. Other accidents are caused by human behavioral problems, such as tardiness, failure to use designated passageways, careless disposal of cigarettes, etc.

Work environments are often seen in which shelves and cabinets are locked. This may very well be intended to deter theft but, ironically, where locks are used or lids closed, the inside of these storage areas is almost always messy.

Although discipline is the last of the 5S's, it is one of cardinal importance.

3.7.4 The Three Pillars Supporting 5S Activities

The key to 5S activities is the one-at-a-time, steadfast implementation of an activity for a specific goal. When viewed as a whole, these activities fall into three categories, which we call the three pillars supporting 5S activities. We will briefly discuss the activities' specific goals and objectives for each of the three pillars (see Table 3.5).

Let us first focus on the disciplined work environment. As shown in Table 3.5, the purpose of this goal is to focus on raising the basic management level through 5S practices. It can be seen as the essence of the 5S's themselves. 5S activities must transform the quality of human behavior. In a nutshell, the goal is to see that everyone abides by the rules once these rules are defined. Thus, of central importance are the ways in which everyone goes about carrying out his or her duties and participates in activities. Of equal importance are the training methods that allow each individual worker to be responsible for his or her assigned activities and behavior. Improvements clearly need to be made in the area of management by observation. Once the 5S's prevail in the work environment and are ingrained in the behavior and activities of individual workers, the next step is to ensure that the acquired habit of disciplined behavior on the part of every employee will not be forgotten or allowed to disappear.

The key to the next step—making a clean work environment—is to clean the smallest elements of the work environment and equipment that have either never been cleaned or have never been thoroughly cleaned and to get rid of all the grime and dust. By thus radically transforming the condition of the work environment, the workers will be both motivated and inspired to renewed awareness and behavior.

The third pillar of 5S is the creation of a work environment that can be managed by seeing, an idea popularly referred to as management by observation. By improving upon ways to identify abnormal conditions quickly and easily, it becomes possible to create an environment in which anyone in an area can spontaneously help someone else when an abnormal condition occurs. This means that measures are taken to prevent people from making mistakes and committing errors, and can be called the standardization of the 5S's.

These activities that comprise the three pillars should be pursued simultaneously to take advantage of the mutual interrelationship among them.

3.7.5 Measures to Be Taken for Management by Observation

People do not always find it possible to act immediately or to adhere strictly to a set of given rules. It was mentioned earlier that the practice of the 5S's is "for everyone to abide by the rules once these rules have been defined." In reality, however, these rules are not always

TABLE 3.5 The Three Pillars Supporting 5S Activities

3 pillars	Creation of disciplined work environment	Creation of clean work environment	Creation of work environment conducive to management by observation
Goals	Raise the management level • Yardstick to test if everyone adheres to promulgated rules	Clean the equipment and work environment • Change people's awareness and create a work environment where every part of the equipment and work environment is well cared for and cleaned	Measures to be taken for prevention of errors • Devise measures so that errors can be easily seen and quickly corrected • Standardization of the 5S's
Specific themes for activities	1. Company-wide one-minute practice of the 5S's (3-minute 5S's, 10-minute 5S's, etc.) 2. Individualized task assignments 3. Straight line and right-angle movements 4. Quick and reactive correction behaviors 5. No locks, lids, or covers 6. 30 seconds for retrieval and restorage 7. Oasis or hub-oriented movements 8. Announcements for synchronized cleaning by workers 9. Single-item inventory issues 10. Orchestrated calisthenics 11. Campaign for use of safety shoes and hard hats 12. Management of commonly shared areas 13. Collection of discarded cigarettes 14. One week of zero absenteeism campaign 15. Sense of my PM responsibilities 16. Planned schedule of actions 17. Training for coping with abnormal conditions: accidents, earthquakes 18.	1. All-out cleanup of unnecessary items 2. Measures against sources of wetness 3. Management of colors (color conditioning) 4. Major cleanup 5. "Spit and shine" campaign 6. Weed removal from external surroundings 7. Covers against wetness and splashes or splatters 8. Improvements in cleaning tools 9. Campaign for neatness and tidiness 10. Campaign for raising visibility ("transparentizing") 11. Improvement in legibility and clarity of writing 12. Notices, bulletins, and signs 13. Prevention of noise and vibrations 14. Improvements in beautification of environment 15. Coordinated campaign for "wipeoff" cleaning	1. Matching symbols 2. Meter-controlled zones 3. Thermosensitive labels 4. Directional and rotational indicators 5. Belt size differentiation 6. Indication of open and closed states 7. Indication of voltage 8. Inspection seals 9. Color-keyed tube plumbing 10. Lubricant specification labels 11. Wiring management 12. Safety indicator color(s) and danger warning color(s) 13. Fire extinguishers 14. Foolproofing indicators 15. Control limit indicators 16. Labels indicating responsible personnel 17. Demarcation lines: passageways, sectional partitioning, clearance zones, corners 18. Places for storage: tools, tips and blades, measuring instruments and equipment 19. Indication of various limits of control 20. Filing 21. .....

132

clear in many cases. In order for people to abide by the rules, it is necessary that the means be provided to create an environment conducive to their observance. For example, in raising a child with a sense of discipline, although parents might tell a child to put his shoes neatly in the entryway closet, the child might not listen to his parents. However, in one instance, when a parent drew a symbol of shoes near the closet, the child's curiosity was aroused, he was interested in responding to the symbol, and began not only to store his own shoes but other people's shoes as well. Traffic signals and road signs are designed to be noticed by drivers so that they can react to them quickly.

Thus, to influence people to abide by rules, it seems apparent that if tools are provided that help people to obey them more easily, or even that force conformance, e.g., if "objects" or things are provided as vehicles for carrying out the desired actions, then people would more readily and intentionally pursue these actions. These vehicles or "crutches" are objects that make it easy for people to measure how deviant they are from the norm, and hence to judge if their conduct is to be praised or condemned.

As tools, we include those items ranging from something on which to place some "things" to objects upon which things can be hung, such as bulletin boards marked with symbols. Table 3.6 shows the criteria and key elements of objects to be used for management by observation.

TABLE 3.6

Tools for management by observation
1. They can be identified from a distance
2. They are displayed on the site or on things that need to be managed
3. They clearly display whether things are in good or bad condition and the differences are immediately apparent
4. They are easy to use by anyone and everyone
5. They help everyone to stick to the rules and correct deviations
6. Their use makes the work environment clean, neat, and tidy

KEY POINTS IN MANAGEMENT

Everything has a key that must be closely scrutinized. For example, if we look at an F-86 fighter plane, we notice that there are various kinds of markings, color-coded symbols, and numbers on the fuselage or in the interior. These may be markers denoting the fuel intake nozzle openings, the pedal positions, color-keyed plumbing works, matching markers for rivets or bolts, the torque value reading fastening, etc. These are some indications of how a fighter plane is designed to allow rapid but error-free maintenance in case of emergencies. When people get physically tired, they may massage a particular spot in a focused manner, or when they get sick, a doctor may examine the tongue or check the pupils of the eyes. These spots are particular areas that may be weak and vulnerable, that may show the early symptoms of illness, and that must always be in a particularly good condition, or they require attention. These areas are the key spots for management. However, they cannot be readily identified by everyone. For machines, these spots are often concealed by covers, and internal key spots are hidden. Thus measures should be taken to make them observable to enable early recognition and identification of problems.

In reality, however, too many manufacturing companies have done little to implement this basic requirement for management by seeing. In many cases we wonder what their managers are really seeing on a day-to-day basis as they go about "managing." By and large, compared to other areas of concern, measures and displays to promote safety are relatively advanced. In many places yellow or red color markers are used for safety indications, and specific areas requiring caution are clearly marked.

On the other hand, there is often an unfortunate method of using various types of checksheets for maintenance activities. These are used in such a way that wherever checkmarks are given, they are relied upon. If a checkmark is missing, however, the viewer might not respect or believe in the validity of the workers' activities. What is needed is not a method that gives credence to a checksheet, but one that would allow us to give credence to the actions of people.

To manage the work environment successfully, the key areas and points of management must first be clearly defined and measures must be taken to make management by observation easier, leading to attentive care and watchfulness in all key areas. Figure 3.28 shows a case in which a manual was created to promote management by observation.

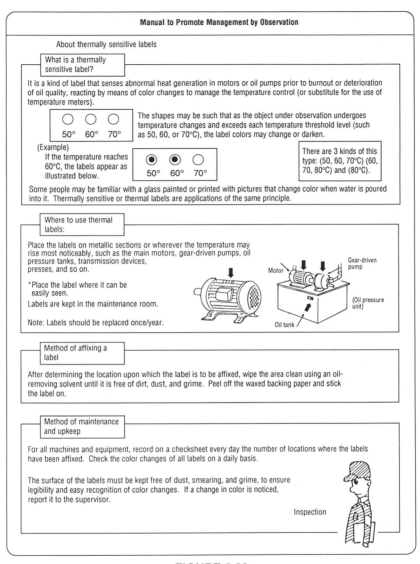

FIGURE 3.28

3.7.6 Dealing With Sources of Dirtiness (the 5S Techniques)

Considerable research has been done on measures to be taken against dirt, burrs, and foreign matter on products, spawning numerous techniques for removal of burrs and dirt and for sealing. Recent developments in electronic engineering have made it increasingly necessary to keep not only the production equipment clean, but also the work environment as a whole. Today there are manufacturing plants that have made cleanliness a special concern.

Measures against dirt mean cleaning, in which all the workers should be involved. In reality, however, it is easier said than done.

What is meant when we speak of dirt? What are the problems caused by dirt? What are the acceptable limits? When we say something is dirty, we are referring to the state resulting from whatever was leaked, generated, fallen, mixed, or blown in. Dirty conditions may fall into different types: a type that ordinarily occurs during manufacturing processes, a type that should never occur, a type of dirtiness that becomes noticeable as a result of repeated neglect of cleaning, a type that resulted from more dirt than workers could manage during the entire cleaning process, etc. It is necessary to investigate how dirtiness occurs and what can be done about it, following a procedure enabling efficient inquiry.

- Does the suction intake nozzle have the right kind of shape and is it pointed in the right direction?
- Are the shapes of the sawdust and flakes subject to easy collection efforts?
- Are the packing and sealing methods adequate as they are used in a dusty work environment?
- Are the shapes of the machines themselves and their ejection chutes appropriate?
- What should be done to shield the machines and equipment from splashes and splatters of oil and water?

This list of questions is theoretically endless, and it shows the variety of subjects to be dealt with in the area of engineering. These vary in complexity from a type that lends itself to individually initiated improvement efforts to a type that resists any fundamental solution unless dealt with using technologies in an organized way.

In order to devise techniques to prevent equipment from becoming dirty (techniques for the 5S's), where and how the equipment gets dirty and what kind of measures are being taken must be understood in a quantitative manner. It is also necessary to deal with these in terms of the specific and unique characteristics of the machines, processes, and products involved.

TABLE 3.7 Types of Leakage and Dirtiness

Type \ Symptoms	Oil	Steam/ Water	Air/ Gas	Sawdust/ Burrs/ Flakes	Sand/ Dust	Shot beads	Oil mists	Fabrica- tion materials	Explanations of the patterns of leakage and dirtiness
① Oil splatters and splashes	O	O		O	O	O	O	O	Oil splatters and splashes that leak out through gaps and holes
② Overflows and spillovers	O	O	O	O	O	O		O	Containers may be overfilled and the contents spill over
③ Adhesion and traveling	O	O		O	O	O		O	Adheres to people, vehicles, and totes or workpieces
④ Spillage	O	O		O	O	O		O	Spillage occurs when resupply, cleaning, or washing is being done
⑤ Run	O	O					O		Liquid-type materials run over something or smudge
⑥ Blotting/ Spreading	O	O	O				O		Liquids blot through small gaps
⑦ Dripping/ Dropping	O	O						O	Parts may drop off their totes or conveyors

INVESTIGATION OF THE SOURCES OF DIRTINESS

To investigate the sources of dirt, a set of procedures should be established, and these procedures must then be followed one step at a time. It is also necessary to focus the investigation by defining what type, shapes, and patterns of dirtiness exist. This point is illustrated in Table 3.7.

After a specific subject of investigation has been chosen, it is necessary to identify its origin, quantity, how it should be measured, and what current actions are taken to prevent it. Figure 3.29 shows an example of the mapping relationships among the elements of investigation.

INVESTIGATION OF PLANNED MEASURES

Figure 3.30 shows two different methods for measures to be taken against dirtiness.

The first method is to conceive of the means to prevent the dirtiness from occurring. This means either completely eliminating the sources of dirt or reducing the amount of dirt generated to an extremely small amount. Both goals require methods to avoid the pro-

Sand leakage MAP

Name of production line (CSS-6)
Casting section 2, operation 23, step 24
Director | Asst. Dir. | Manager | Supervisor
Date of investigation: November 24, 1990
Name of investigator: J. Smith
Section serial no. / Line serial no. — 1/2

Leakage Map ID no.	Machine unit no.	Component	Splash	Overflow spills	Runs and drips	Stuck frames	Volume observed (Dipper no. of fillups/day)	Status of measures	Number of recovery operations	Receptor box Yes	Receptor box No	Measures for improving recovery rates (tools)	Ranking order of preference	Repair	Improvement	New installment	In-house manufacture	Order placed to maint.	Order placed to mfg.	Cost	Started	Completed	Binary assessment O or x
1	BC-103	Belt		✓			5		0.1 hr/day	✓		Shovel/broom	27		O				O				
2	BC-104	Top screw		✓	✓		7	Receptor chute/duct	0.25 hr/day	✓		Shovel	16		O				O				
3	BC-105	Belt		✓	✓		2	·	0.25 hr/day	✓			17		O				O				
4	FS-5	Balancer valve	✓				2 cups/day	None		✓			28		O				O				
5		Anvil cleaning	✓				2 cups/day	·		✓			22		O				O				
6	F-10	Molding mounting device			✓ (Attachment of ballast)		5 cups/day	·	10 hr/day	✓			23		O				O				
7	PE-40	Empty upper frame PS				✓	3 cups/day	Leaks when transferred to measuring tote/head assembly		✓			29		O				O				
8	MO-90	Upper mold		✓			10 cups/day				✓	Being recovered	30		O				O				
9	BC-139	Heavy compressor (top)		✓	✓		10 cups/day	Skirt	0.25 hr/day		✓	Widen belt width	8	O				O					
10	BC-139	BC 140 intersection		✓	✓		20 cups/day	Scraper	0.5 hr/day		✓	Prevent inertial moves	4		O				O				
11	"	BC138 intersection		✓	✓		30 cups/day	·	1 hr/day		✓	Prevent inertial moves	1		O				O				

Column note: "Record the no. of fillups/day" (Volume observed); "Record the status" (Status of measures); "Record the no. of operations/day or week" (Number of recovery operations); "Record the tools used to cope with problems like • Bail-out tools • Cleaning machines" (Measures for improving recovery rates).

FIGURE 3.29

138

Ideas	Specific measures (5S Techniques)	Improvement
1. Methods of preventing sources of dirtiness 1) Prevent them 2) Reduce occurrences	Prevention of leakage: tight enclosure methods, sealing methods	① Remove
	Prevention of splashes and splatters: shapes of doors and covers, directions and shapes of splashes	② Wipe
	Prevention of drops: conveyance methods, injection methods, shape of containers	③ Fix
	Repairs of loosening and breakage damages	④ Quit
	Investigation of operations: no burrs, oil-free manufacturing, no polishing requirements	⑤ Stop
	Prevention of stuffing and sticking of machining debris	⑥ Reduce
2. Methods of collecting and removing dirt 1) Methods of collection 2) Methods of removal	Debris collection capacity, review of methods used Capacities and shapes of the suction intake ducts	⑦ Do not accumulate
	Methods of removal and withdrawal Cleaning tools, recovery ducts, shape and size of the receptor bins	⑧ Collect
	Washing methods	⑨ Do not spill
	Shapes of debris, size and direction of scattering	⑩ Do not carry around
	Shape of the platform or main body of the machines	⑪ Throw away

FIGURE 3.30 Method for Organizing Ideas on Measures Against Sources of Dirtiness and for Advancing the Implementation Plan

duction of dirt. Some measures may require various kinds of investigations that are of an engineering nature. Where machines are replacing human manual labor, damaged areas must be repaired and their daily hardware maintenance constitutes an important part of these activities.

The second method considers efficient methods of collecting and removing dirt. Since some manufacturing work may always involve processes that generate dirt, the problem is both constant and familiar. These methods include a central collection system such as drainage ducts for collecting oil or waste water, or ducts and vents to collect and prevent the scattering of dust, a method that focuses on tools for easy cleaning, or the design of machines to allow smooth drainage of debris and easy cleaning.

Figure 3.30 illustrates key improvement points in investigating measures to be taken. When a specific set of planned measures emerges, it must be evaluated on the basis of discriminating criteria. These criteria may be based on such considerations as the estimated costs or the number of operations needed for the new proposed measures, how easily the measures can be accomplished, and whether the engineering aspects of the measures need to be investigated by other departments before any conclusion can be drawn. When implementing the plans, one should have an estimate of the expected results to be derived from the planned measures, and rank the results in order of significance.

Central to the measures to be taken are constant improvement and daily maintenance of the operations and equipment that are the sources of dirt and leakage. The causes of their dirtiness and leakage should be understood and engineering remedies should be investigated as well. Ironically, in many plants people are not very good at controlling dirt and leakage, despite the fact that the technique and know-how of preventing dirt and leakage have been in extremely high demand. The specific engineering requirements for controlling dirt and leakage may vary. It is extremely important, however, for manufacturing enterprises to organize the knowledge and expertise on these topics and to "own" them internally as their own engineering savvy.

3.7.7 Advancement of the 5S's

1. THE 5S PROMOTION PLAN

There are more topics concerning 5S activities, too numerous to be listed here. Implementation of any of them may potentially lead to success in the attempts to establish a perfect control method. In pursuing 5S activities, it is crucial to deal exhaustively with one task at a time, even though the item in question may appear trivial.

5S activities can generally be categorized as follows:

	Examples
(1) Make rules and abide by them . . .	Eliminate unneeded items, major cleanup, 3-minute cleaning
(2) Create tools, use them on . . .	Cabinets, shelves, various display labels, markings, dirt, and leakage
(3) Must-improve, must-modify items . . .	Antisplash, antispatter covers, various antileakage measures
(4) Items that must be ordered for . . .	Removal of unneeded machines, equipment, other departments, plumbing; oil leakages

Categories (1) and (2) require intensive human labor, and unless all-out efforts are made at deploying a large number of people, the results may be insignificant. These efforts must be consistently maintained through practice, otherwise they may be forgotten or gradually neglected.

For categories (3) and (4) considerable ingenuity is needed, as well as investigation of methods, time factors, and costs. None of these are as simple to implement as they may appear to be. A key to success is the clear identification of what can and cannot be accomplished by each designated group.

Figure 3.31 illustrates the key process of establishing a basic plan for 5S advancement. The tasks should be divided into several steps, the first being a major cleanup effort to eliminate unneeded items. A manual should be published that can be used as the foundation for a step-by-step implementation of the program. Measures to be taken against the sources of problems, secrets of successful management by observation, and improved management tools are typical items to be discussed.

2. THE 5S MANUALS

Despite the universality of the problems, 5S practices are not deeply rooted. One of the reasons for this situation is the lack of clearly

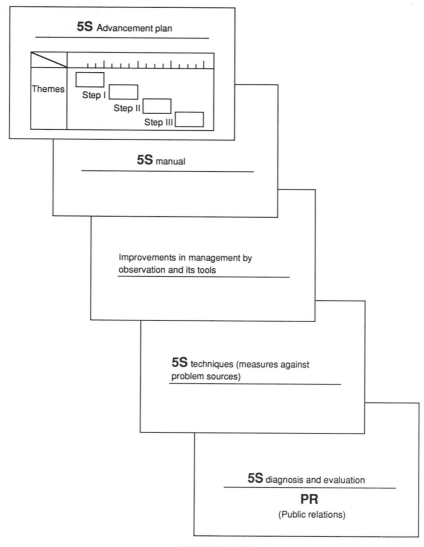

FIGURE 3.31

defined standards from which to determine what should be done and for executing the 5S plans.

3.7.8 5S Diagnosis and Evaluation

The diagnosis of 5S practice and its evaluation deals with the question as to whether 5S activities have truly been progressing, becoming part of people's behavior, and producing consistent results. In theory, so long as everyone puts his or her mind to it, obstacles or problems should not occur. In reality, however, people cannot set aside a solid segment of time to practice in a concentrated manner. Rather, everyone must practice while busy trying to perform regular routine duties. Thus, unless there are goals that are commonly agreed upon about the pace and level of perfection, it is next to impossible to achieve concerted efforts among the people involved.

1. DIAGNOSTIC METHODS

There are two main methods that can be used:

1. Level method
2. Competition method: League or "round-robin" competition method and tournament method (as in a championship tennis tournament)

The "level" method consists of observing a designated place, comparing it against a set of standards, and judging which level of accomplishment has been attained for that location. The judging may be done either by a self-evaluation method or by having an authorized judge or third party to examine the case at hand.

The competition method is based on comparison among a set of competing teams responsible for their designated locations. It is important that the competitive levels be more or less equally matched to preserve the game aspects of the competition. The league competition method lets each team compete with any and all of the teams in the group, but individual competitions take place on a one-on-one basis. At the end, the win-loss counts are tallied and the resulting scores are used to determine the final evaluation. The tournament method involves a process of elimination whereby each victory entitles a team to go up a level in the tournament "tree."

When the level of each team involved is relatively high, the competition method is both interesting and self-perpetuating since it preserves a competitive spirit among the teams.

For either the "level" or competition method, meaningful diagnosis depends on devising ways that motivate people to practice. Without motivation, performing the evaluation becomes pointless.

2. DIAGNOSTIC SYSTEM

In the example of a diagnostic system discussed here, diagnostic criteria are set up for each level of advancement in practice. First, the personnel in a particular work environment conduct a self-initiated diagnosis. If their environment meets the criteria and passes the test, then the official diagnosticians will give a certificate of achievement for that level. If a group passes the certification test for any given level, then at a later time, a surprise inspection and diagnosis is carried out and the group is judged again. This mode of inspection is applied to all groups at each level.

3. EVALUATION STANDARDS

Standards must be set to determine the winner and losers, or whether a group passed the test for a particular level. The official diagnosticians and the judges should not only be the ones who are aware of the standards. For instance, people often feel confident about their understanding of a textbook's contents, but when tested, they realize

TABLE 3.8 An Example of the 5S Diagnostics Checksheet

Category		Items	Manual	Findings	Remarks
Floor surface (stepping platform)	1	No oil or debris is spilled on the surface.			
	2	No dirt or parts are scattered on the surface.			
	3	No scraps are left on the surface.			
	4	The surface is not dirty.			
	5	No floor-level obstacles are on the surface, such as protrusions, cavities, cracks, or paint flakes.			
	6	No dirt smears, damage, scraps, or flakes are on the demarcation lines and positioning marks.			
Pallet totes Manual lifts	7	The name of the person in charge is displayed on the pallet totes and the manual lift.			
	8	There are no damaged sections in the pallet totes or manual lift.			
	9	The wheels of the pallet totes and the hand lift are in a normal condition and there is no dirt or debris.			
	10	There is a display indicating where to store the pallet totes and manual lift.			
Totes Parts bins Cardboard boxes	11	The totes are placed in designated positions, along straight or perpendicular lines.			
	12	There are no elevation-positioning wood pieces used in the totes, and the height of the totes is not above the standard level.			
	13	There are no damaged areas in the totes.			
	14	There is no dirt or debris found on or in the totes.			

that they did not fully understand the material. Similarly, people must know the evaluation standards, using them to conduct self-initiated diagnoses, discover their weaknesses, and be given the opportunity to remedy them. If this information and opportunity are not provided, the reaction to a test might be "Well, go ahead, we couldn't care less; it's just a matter of luck anyway." This epitomizes a counterproductive symptom.

Table 3.8 shows an example of the diagnostic checksheet. If diagnoses using a checksheet are done for each level, then the checksheet that is used first should be based on a simple method.

4

Planning and Managing Maintenance

4.1 A SYSTEM OF PRODUCTIVE MAINTENANCE AND MAINTENANCE DEPARTMENT TASKS

Production activity goals are expressed in terms of productivity improvement (P), quality assurance (Q), cost reduction (C), delivery date adherence (D), safety and environmental protection (S & E), and improving worker motivation (M). Productive maintenance activities are performed to achieve these goals, a systemic view of which is illustrated in Figure 4.1. Understanding this systematic view requires research into the reliability of equipment to improve their effective utilization, research into the maintainability and supportability of the equipment, and a lifelong commitment to the investigation of equipment costs. Specific analyses and maintenance activities should be carried out for these investigations.

To promote PM activities, the maintenance department must first improve its efficiency and level of expertise. Figure 4.2 presents a systematic view of all the tasks facing a typical maintenance department.

The major tasks can be organized into four categories. First, maintenance department activities must be made efficient in terms of planning, standardization, and flexibility. These topics will be dealt with later in regard to the maintenance calendar.

As shown in Figure 4.2, readiness and versatility are the characteristics most often demanded of the maintenance department. Repairing sudden breakdowns is a function of the maintenance crew; these crews function much like an ambulance crew or fire engine

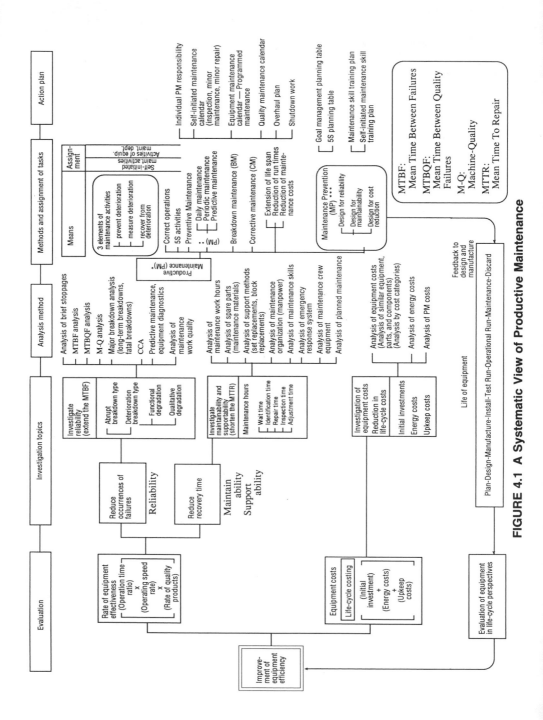

FIGURE 4.1 A Systematic View of Productive Maintenance

Definitions for Figure 4.1

*Productive Maintenance (PM)--Seisan hozen: Equipment maintenance for enhancing the profitability of production by decreasing the sum of the cost of equipment itself, as well as all the costs involved in the maintenance of equipment and the degradation loss of equipment throughout the entire process of planning, design and production, operation, maintenance, and scrapping of the equipment. It includes the functions of breakdown maintenance, preventive maintenance, corrective maintenance, and maintenance prevention.

**Preventive Maintenance (PM)--Yobo hozen: Equipment maintenance to prevent the occurrence of abnormalities through intentional checking and arrangement, adjustment, oiling, and cleaning. The equipment is constantly maintained in a normal and good condition while trying to perform such maintenance work on a paying basis.

***Maintenance Prevention (MP)--Hozen yobo: An equipment maintenance activity aimed at the establishment of a system that will make equipment maintenance almost unnecessary through the selection, supply, or design of equipment having excellent operability, reliability, maintainability, safety, productivity, and the like, based on information and novel techniques for equipment maintenance.

company. In reality, however, many maintenance departments are not as responsive and versatile as they should be, and considerable efforts should be made for improvement. Some of these efforts will be discussed in the section dealing with improvements in maintainability.

Hardware preparedness, also referred to in Figure 4.2, points up the need for the maintenance crew to be sufficiently equipped to cope with equipment problems. Just as aircraft and tanks are used in battle, the maintenance crew must deploy a variety of efficient and up-to-date maintenance tools. The use of new diagnostic technologies for diagnosis and computers for maintenance information processing, equipment failure analysis, and maintenance planning should increase significantly, a role for which aggressive management attitudes are needed. At the same time, the contents of maintenance activities should gradually undergo a transition, as shown in Figure 4.3.

The second task is to increase the level of expertise by analyzing the engineering aspects. To carry out maintenance with a high level of expertise, the method of analyzing failures should be improved, and, subsequently, the reliability and maintainability of the equipment.

Maintenance records are the first step toward this effort. A review of these maintenance records reveals a great deal about the level of maintenance expertise. In many plants, however, the records are seldom used appropriately, as was discussed in the sections dealing with MTBF and MTBQF.

The third task is to raise the level of maintenance skills and personnel, including the capabilities of individual members of the main-

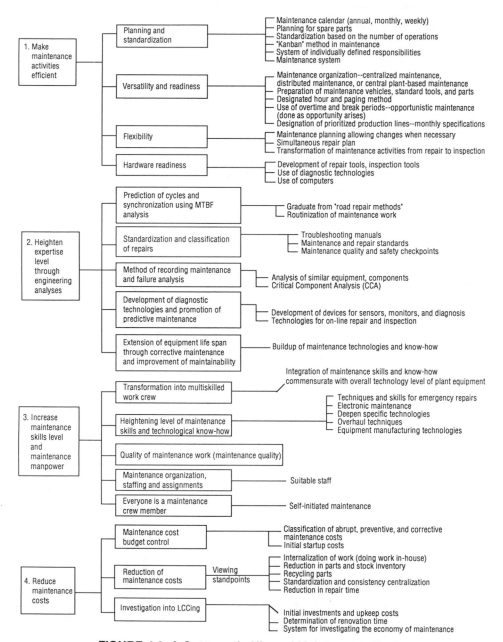

FIGURE 4.2 A Systematic View of Maintenance Tasks

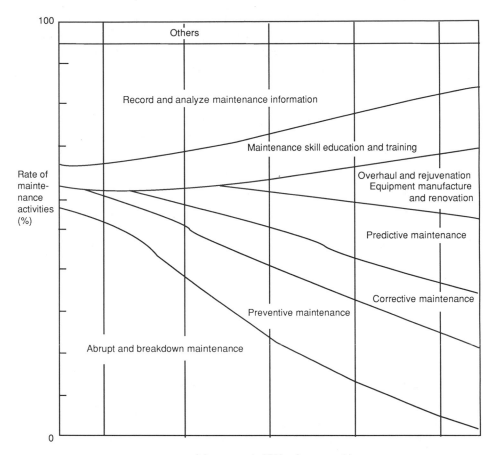

FIGURE 4.3

tenance crew who are confronted by the increasing complexity and sophistication of engineering and technology. Just as most people to-day cannot adjust their automobiles and television sets, similar situations exist and are becoming more prevalent on factory floors. Despite the increased level of demands for emergency repair and overhaul engineering skills, the supply of such skills has been declining.

Another aspect is an increase in the type of maintenance personnel available, a result of the spreading practice of promoting programs for self-initiated maintenance and for multiskill-oriented education and training.

The maintenance engineering technology of the future may very well be transformed into a "technology for replacing parts," as amply exemplified in electronics. Among maintenance engineering technologies, prediction of abnormalities and diagnostic technology may play significant roles. Maintenance education and training programs should seriously consider these aspects in the planning stage.

The fourth task is the reduction of maintenance costs. Currently, except for large-scale equipment and machine-dependent processing industries, sufficient attention is not being given to maintenance costs. Nevertheless, the absolute level of maintenance costs remains high.

A need clearly exists to break down maintenance costs according to goals, and to raise the level of management expertise in budgeting maintenance costs.

With regard to maintenance cost reduction, many cases occur whereby these expenses are decreased following a manager's snap decision under the slogan "Let's cut the maintenance costs." But a more appropriate way to handle this is to know the maintenance budget during the budget management cycle, and carry out budget cuts based on reason. This subject will be dealt with in greater detail in discussing spare parts inventory management (see section 4.7).

4.2 KEY CONSIDERATIONS IN DESIGNING THE MAINTENANCE SYSTEM

4.2.1 Factors in Creating a Practical and Useful System

The rapid pace of change in the contemporary manufacturing and processing industry environment calls for an innovative response that freely jettisons some of the old customs and institutions that are no longer relevant. The PM system is particularly aware of the need for plants to cope with rapid change.

The following is a discussion of several items pertaining to the fundamentals of maintenance systematization, which are the basis for designing a PM system.

1. SUFFICIENT WEIGHT SHOULD BE GIVEN TO SIMPLIFYING THE ELEMENTS OF THE SYSTEM.

The successful round trip of the Apollo moon rocket required the complex, consecutive interaction of many parts and the overcoming of multiple problems. When we think of such a large and complex system and its supporting technologies, there is a tendency to fall into the trap of thinking that systematization means making a complex system consisting of many parts. On the contrary, systematization should be thought of as a technology for making complex things into a simple whole.

2. THE RELIABILITY OF COMPONENTS AND OF THE WHOLE SYSTEM

The next step emphasizes the importance of the individual and component functions that together comprise the system as a whole. At first, this may appear diametrically opposed to the philosophy that emphasizes the idea of a total system. This, however, is not so. The intent here is to emphasize the need for each individual component to be on a solid footing, thus avoiding anxiety about the stability of a system comprised of individual components. This does not imply autonomous behavior by individual components. Rather, each component is firmly rooted in a foundation of basic units. The combination of such firmly rooted components can create an extremely sturdy whole.

This thinking applies to a number of other areas, and it helps clarify overall management problems, the problems of life cycle PM maintenance management for equipment, questions about the whole versus individuals, and the total system versus subsystems.

3. SYSTEM DESIGN AND DETERMINATION OF ITS OBJECTIVES

A common trap may ensnare people who are well-versed in various aspects of systems and system engineering. They might make an exhaustive analysis of the system requirements confronting them, and then design and develop a system or systems. The catch, however, is that their systems are not responsive to the very objectives for which they were designed. From the administrative and management point of view, these people tend to operate systems without a clear direction—or, where clear directions are given, directions for action are often abstract, holistic, and lack relevance to specific individual management objectives.

When an Apollo rocket was launched, a specific objective existed: to reach the moon and return. Likewise, system design and development should have specific and clear-cut objectives. If high goals are set, measures should be taken to cope with the situation by bridging the gap between the aim and its accomplishment. On the other hand, if the objectives can be accomplished without making efforts to narrow this gap, we can say that the means chosen may not have to withstand critical scrutiny, and they may not even need to be managed. Depending on the kind of objectives to be accomplished, widely differing methods need to be applied.

4. WHEN A SYSTEM FAILS TO FUNCTION

When a system fails to function as desired, it is often because there was no investigation or analysis of what the system really needed to do or what it was ultimately to provide.

Two aspects should be focused on here: (1) short-term goals for PM, i.e., for reducing current period costs; and (2) the long-term perspective, i.e., ambiguities regarding the future PM promotion program for achieving qualitative improvements in the plant. In the latter case, the system design should take into consideration from the outset the ultimate state of the plant in which it will operate.

5. THE SYSTEM AND ITS ACTION STANDARDS

If we view a system as a kind of theatrical stage on which actions or groups of actions are taken, a set of action standards must be set to keep the whole plant moving along. Sometimes, however, such standards are found to be merely theoretical, or they are often too cumbersome and convoluted to be well understood. This indicates the absence of a consistent yardstick for evaluating the action unit as a whole, which in turn causes instability in its momentum. It is therefore necessary to identify, define, and adhere to the roles and values of assigned tasks for each individual managerial unit.

6. THE KEY TO AN EFFICIENT SYSTEM

There are variations to this rule, depending upon the characteristics of various enterprises. The efficiency of system-related activities, however, is inextricably intertwined with the functions of the enterprise and with the functions and roles of individual subsidiary organizational units within each business enterprise. The situation is similar to the role of electrical control circuits in the design of machinery, equipment, or mechanical systems for machines. The difference with management systems is that the elements of which they are comprised are not mechanical parts or electrical components, but people. Therefore, even if the standards for actions to be taken within the framework of the systems are clearly defined, the act of judgment determining behavior is made by human beings. This means that when there is a need to act in response to dynamically changing specific problems that may undergo an unpredictable number of permutations, the final choices of action are determined by human will. Understanding this is the indispensable and necessary condition for the system to make any goal-oriented inferences or be aware of actions leading to successful management. Ultimately, this means simply the problems of raising the overall level of awareness and discipline, creating the motivation to be self-starting and thinking in terms of the systems' perspectives.

It is important that people perceive themselves in a holistic manner with respect to that which precedes and follows their actions. This includes, for example, awareness of the relationship between the individual's work environment and the total environment; awareness of the extent to which one's colleagues give credence to the reliability of each other's work; what kind of organization is effective with regard to communication with one's peers, subordinates, or other departments; what kind of impact or effect will the making of a particular procedural move have on others. These factors pertain to credibility and to goal directedness and awareness. In sum, a revolution in human awareness must first be achieved before efforts at systemization can be promoted. To do otherwise is to create more problems than solutions. The problems pertaining to systematization and human revolution are essentially inseparable.

Efforts to advance systematization must be made through the collaboration of all plant departments. There must be clearly defined institutional and organizational arrangements to promote cooperation among the staff, project leaders, and line workers on the floor. It must be emphasized that establishing cooperative organizational arrangements is the key to successfully advancing systematization and to acculturating the enterprise in a systematized manufacturing method. To achieve this, people's awareness is focused on a specific set of goals. At the same time, various job functions within the enterprise are also given a goal to focus on.

4.2.2 Considerations in Designing an Organization for PM

In planning and designing PM advancement systems, no set patterns exist that are suited to all business enterprises. The following eight points, therefore, should be considered when planning an organization for PM. It should be noted that business enterprises are always in a state of flux of development and expansion, and that surrounding circumstances are also changing. Organization for PM advancement must thus adjust to and accommodate these changes.

1. Product characteristics: raw materials, semifinished goods, physical, chemical, and economic characteristics of the finished goods.

2. Production modality: processing, fabrication and assembly, continuous processing, number of shifts.
3. Characteristics of the equipment: extent of automation and modernity of the equipment, speed of structural and functional depreciation, degree of depreciation.
4. Geographical conditions: conditions of the business environment, extent of concentration or dispersion.
5. Size of plant.
6. Makeup and background of human resources: level of skills, management levels, human relations.
7. Extent of subcontracting: ease or difficulty of use of subcontracting capacity and its economy.
8. Equipment management in line with processing industries.

4.2.3 Design and Operation of the Maintenance System

PM activities should be pursued in such a way that each of the subgoals is organically linked to every other. Unless these activities are integrated and practiced in daily routine management methods as part of the ordinary "business as usual" mode—whether they are planning, implementation, evaluation, or reactive measures—they cannot really be called PM activities.

Figure 4.4 presents an example of a system of maintenance activities centered around self-initiated maintenance department activities. On the left side is shown a system based on a maintenance calendar of the equipment maintenance department. This section illustrates the flow of activities originating from the annual maintenance calendar that links to the implementation and review of monthly and weekly maintenance activities. Connected to this are spare parts and maintenance activity management. The right side illustrates maintenance activities that are self-initiated by the departments using the equipment. This segment is linked to the equipment maintenance department by means of the MTBF analysis and maintenance review committee meetings. The self-initiated maintenance activities are also linked to lubrication maintenance activities through activities of daily lubrication, the 5S's, management by observation, and activities for reducing brief stoppages. Maintenance activities are recorded in the maintenance report, and, through the MP information provided in the MTBF analysis, the contents of the report are reflected in measures against breakdowns and planning for the next generation of equipment.

An example of the maintenance report is shown in Figure 4.5. Self-initiated maintenance activities focus primarily on preventing deterioration of equipment and deterioration measurement. The maintenance department activities, on the other hand, consist of measurement of equipment and tool deterioration and activities aimed at recovery from deterioration. Also included in this category are data collection and analysis of abrupt equipment failures. Computers are used to process the maintenance data.

4.3 ANALYSIS OF EQUIPMENT EFFICIENCY AND DEVELOPMENT OF A PLAN FOR IMPROVEMENT

While the basic goal of successful PM promotion by creating a PM-directed "mind-set" on the part of all plant personnel is being pursued, we must also investigate the nexus of problems contributing to a lowered efficiency level of the equipment and take appropriate measures. Procedural steps for analyzing and improving the equipment's productivity level fall into two separate categories: (1) investigation of the equipment from the standpoint of improving its reliability level; and (2) investigation of activities to improve maintenance activity efficiency. A project team should be assigned to each category of investigative work.

(1)	Improvement in Reliability	(2)	Improvement in Maintainability	
	Relentlessly reduce maintenance needs arising from the equipment (Make efforts to prevent them from breaking down.)	×	Efficiently accomplish maintenance work of the equipment (Fix immediately when a breakdown occurs.)	= Availability
	Investigation into equipment		Investigation into work activities	

The results of PM are expressed in terms of the product of (1) × (2), called the "availability." Availability is used as a measure of expressing the overall level of equipment utilization.

The equipment-related investigation to improve equipment reliability, mentioned before, consists of (1) activities to reduce the number of occurrences of unscheduled maintenance activities necessitated

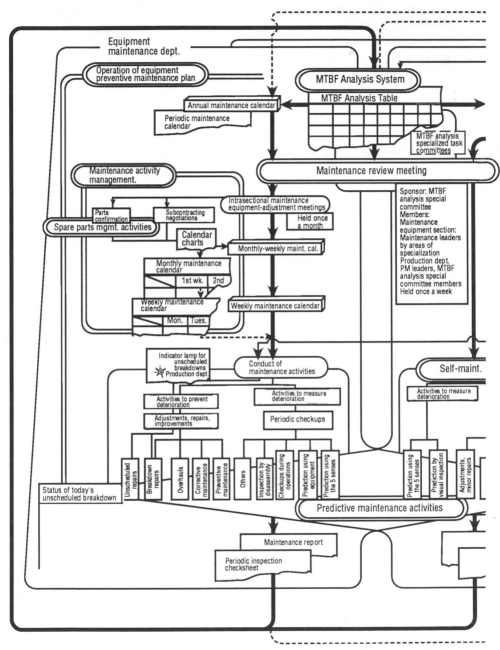

FIGURE 4.4 System for Managing Activities to Prevent Equipment Failures

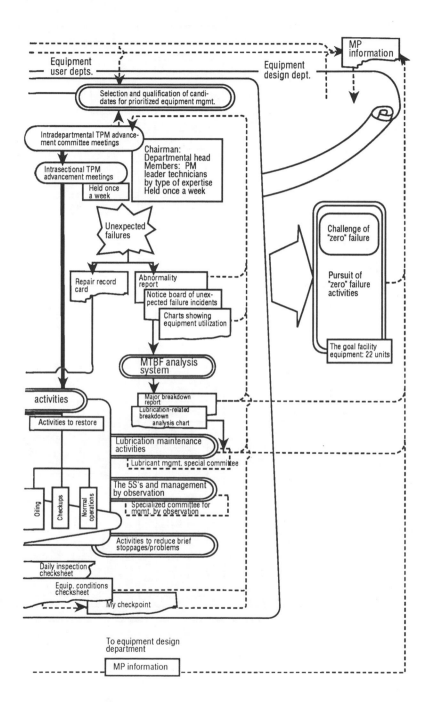

Equipment user depts.

Equipment design dept.

MP information

Selection and qualification of candidates for prioritized equipment mgmt.

Intradepartmental TPM advancement committee meetings

Chairman:
Departmental head
Members: PM
leader technicians
by type of expertise
Held once a week

Intrasectional TPM advancement meetings

Held once a week

Unexpected failures

Challenge of "zero" failure

Pursuit of "zero" failure activities

Repair record card

Abnormality report

Notice board of unexpected failure incidents

Charts showing equipment utilization

The goal facility equipment: 22 units

MTBF analysis system

activities

Major breakdown report

Lubrication-related breakdown analysis chart

Activities to restore

Lubrication maintenance activities

Lubricant mgmt. special committee

Oiling

Checkups

Normal operations

The 5S's and management by observation

Specialized committee for mgmt. by observation

Activities to reduce brief stoppages/problems

Daily inspection checksheet

Equip. conditions checksheet

My checkpoint

To equipment design department

MP information

FIGURE 4.5

Maintenance Report
(Equipment repair request-Report-Review notes)

Reviewer:
Post no.
To:
Section/Unit

*No.
Repair request

*Issued: March 28, 1990, 23:00
*Requested repair date: Mar. 28
Completed: March 29, 02:00

*Dept. no.: 4864
*Machine name: "Dedicated" drilling machine
*Machine no.: DR-915
*Line or product name: Starter yoke
*Issued by: G. Johnson

*Abnormal component or status descriptions: The third unit's drill will not go forward.

Maintenance performed by: R. Delaney — (①) Maint.

Category	Codes / Failure mode
Electrical	Short circuit 11, Broken wiring 12, Grounding 13, Bad contact 14, Bad insulation 15
Frictional	Wear 21, Corrosion 22, Fatigue 23
Positional	Slackening 31, Dislocation 32, Positional shift 33, Warp 34
Physical damage	Break 41, Crack 42, Peel 43, Dirt damage 44
Thermal damage	Burnout 51, Fractured weld 52, Abnormal temp. 53
Pressure damage	Leak 61, Congestion 62, Abnormal pressure 63
Vibration	Abnormal noises 71, Vibration 72
Others	Operational failure 91 (circled), Unknown causes internal & external 98, 99

Activity / Maintenance method:
Preventive maintenance — Checkup, inspection 11, Periodic adjustment 12, Renovative repairs 13, Preventive maintenance 14
Breakdown maintenance — Unscheduled brkdn. repairs 15 (⑳), Breakdown repairs 22, Corrective repairs 31, Equip. compatibility work 41, Safety measures 51, Remodeling, renovations 81, Others 91
Corrected slide shaft's abrasions

Maintenance performed by:
Dept. — Maint. (①), Engineering equipment 2, Materials 3, User dept. 5, Outside source 6, Contracted out 7, In-house work 8, Internal & external 9
Collaboration — Contract out, In-house
Others — Operational failure (circled 1), Location and ID of the repair

In-house work: 5
Unknown causes: 5 2 1

Contents of maintenance work (repaired areas and repair description).

Because of abrasive contacts between the main unit and the slide shaft assembly, the slide shaft became inoperative. The unit was taken apart and inspected, the cause of the abrasion removed, and the installation adjustment made.

Replacement parts

Part	Mfr.	Model	Qty.	Price
				$
				$

Repair Reasons

Design	11	Mechanism
	12	Parts selection
	13	Material or processing
	14	Assembly detail
	19	Others
Manufacture	22	Part fabrication
	22	Processing
	23	Assembly/adjustment
	24	Others
	29	Others
Maintenance	31	Inspection
	32	Adjustment/alignment
	33	Repair
	34	Emergency measure
	35	Maintenance unexecuted
	39	Others
Use	41	Operation
	42	Inspection
	43	Lubrication
	44	Cleaning
Abnormal	51	Abnormal load
	52	Accidental damage
Deterioration	61	Normal
	⑥2	
Improvement	71	Safety
	72	
	81	Prevention of recurrence
	82	
Others	98	Unknown cause
	99	

Maintenance losses / Maint. costs

Maintenance losses	☆ Hours of equipment stoppage	0 hrs.
	☆ Hours of lost production	0 hrs.
	☆ Monetary amt. of defective parts/components	
	☆ Monetary amt. of other losses	
Maint. costs — In-house	In-house man-hours	3 man-hrs.
	Vendor service personnel	man-hrs.
	Subtotal	man-hrs.
	Cost of replaced parts/assembly	$
Subcontracting	Outside contract cost	$
	Outside personnel cost	Personnel hrs.

Interrelationships

Response measures by affected dept.

- Record descriptions of measures, scope, & time spent to prevent recurrence of problem(s).
- Pay particular attention to feedback on machinery and equipment.
- If there will be delays in taking reactive measures, record intermediate results of the schedule/plan for measures to be taken.

Type

1. Needs measure to be taken
2. Reference information
3. ③ No measure taken

Dept. related to:

1. Planning
2. Design
3. Manufacturing
4. Maintenance
5. User
6. Administration
7. Manufacturer
8. Repair service vendor
9. Others

Type of interrelationship

③ — (copy) → Maintenance section (keep)
② — (original) → Impacted section (keep)
① → Prdn. engineering section mgr.

Affected parties (enter the answers)

If an external manufacturer (vendor) is affected, this should first be routed to the maintenance section.

Route

Supervisor of production affected (seal) → Supervisor of maintenance (seal) → Supervisor of production (seal) → Manager of maintenance (seal)
Supervisor of mainte-nance

Notes

* ☆

A. The columns marked with an asterisk * should be entered by the issuing person.

B. The columns marked with a star ☆ should be entered by the production supervisor after the work has been completed. (For hours used, use 0.5 hour as the incremental unit.)

C. With regard to the contents of actual maintenance work and items to be entered as the reactive measures to be taken, the logs of these items may need more space than allotted on this report form. When more space is needed to describe the maintenance, use a separate sheet of paper.

by abrupt equipment failures to zero; (2) the activities for maximum extension of the cycle period of planned maintenance activities (inspection, lubrication, periodic adjustments, etc.) based on investigative engineering work; and (3) activities to reduce the number of operations needed to complete maintenance work that may actually arise. It should be noted that while it is tempting to try to reduce maintenance work to an absolute minimum, this zero maintenance measure may not always be the most economical solution. It is therefore necessary to investigate how to manage maintenance work most economically and efficiently, both in response to problems arising from equipment breakdowns or in a planned maintenance program. If the amount of work generated by equipment breakdowns can be reduced by thirty percent, and the hours needed to absorb the work load can also be reduced by thirty percent, the following formula can be stated:

$$
\begin{array}{|c|}
\hline
\text{(Generated} \\
\text{Work Load)} \\
(1-0.3) \\
\hline
\end{array}
\times
\begin{array}{|c|}
\hline
\text{(Improvement} \\
\text{in Work)} \\
(1-0.3) \\
\hline
\end{array}
=
\begin{array}{|c|}
\hline
\text{(Work Load after} \\
\text{Improvement)} \\
(0.49) \\
\hline
\end{array}
$$

As expressed here, the human resources needed for equipment maintenance and production stoppage can be reduced to one-half.

4.3.1 Investigation of Equipment for Reliability Improvement

Two approaches are used to shed light on the problems and issues of reliability improvement. The first is the machine-quality (M-Q) analysis method (see section 3.3.2), which focuses on the transitional changes in the quality of products produced and the equipment deterioration. The other is the mean time between failures (MTBF) analysis method (see section 3.2.5), which analyzes breakdowns of physical stoppage and functional degradation types. The results of these analyses are then used to define the problems and to determine their characteristic trends. The results are also used to review remedial measures and, after being filtered through a set of routine patterns, to review the action standards of the operation workers.

To use methods that are suited to the particular characteristics of equipment and production methods, efforts must be made to prevent equipment stoppages in proportion to the increases in demands for more production. It is also necessary to take full advantage of the times when equipment is stopped, be it for manufacturing reasons or abrupt breakdowns. Unless these kinds of dedicated efforts are made, sufficient maintenance cannot be accomplished. A carefully planned and well-organized maintenance plan must be implemented by maintaining close ties with the manufacturing department, clearly understanding the realities of the equipment operation plan, making advance plans in preparation for both estimated and planned operation times and for deploying replacement parts. In this way, one can readily judge, for example, if a particular repair of an operation stoppage can be carried out in half an hour or, if the stoppage is for two hours, which combination of maintenance activities should be used.

Without this kind of careful organization, maintenance activities cannot truly be linked with the goals of the production department.

A troubleshooting manual should be created, and all maintenance crew members should thoroughly understand and master its contents. Some elements of the manual should be taught to production workers, so that these personnel can handle certain types of troubleshooting.

Maintenance records should be oriented to the following:

1. In what circumstances did the problem occur or is it occurring?
2. Could it have been discovered sooner?
3. What specifically needs to be done?
4. What improvements can be made to prevent a recurrence of the problem?

This summarizes the study of equipment. Patient and tenacious followup is required for newly instituted maintenance activities; a summary of these measures is shown in Table 4.1.

4.3.2 Study of Maintainability Improvement

The study of maintenance work involves investigation of ways to improve the rate of equipment effectiveness. Using equipment as the medium, it aims to analyze the activities of all workers involved in production activities. One approach may be to identify the scope of work actually taking place within a certain period of time (i.e., a weekly sample), and then to analyze problems and investigate remedial measures within that time frame. Such activities would typically include repair maintenance work, if necessary; checking the level of intensity of work among production workers; checking the conditions of work performed by floor supervisors; checking the work of design and maintenance technicians; etc.

Based on the MTBF Analysis Table, identify those machines with

TABLE 4.1 Measures to Improve Reliability

Improvement of reliability (Reduction in occurrences of breakdowns)

Key points:

1. Ascertain if the breakdown is an initial or chance failure, or if it results from wear and tear.
2. Ascertain if the breakdown is a functional stoppage, functional deterioration, or a deterioration of quality.
3. • Activities to prevent deterioration
   Daily maintenance (inspection, lubrication, cleaning, adjustment, minor tuneup)
   • Activities to measure the extent of deterioration (predictive techniques)
   Equipment inspection (in-operation inspection, disassembly inspection)
   • Activities to restore deteriorated equipment
   Maintenance, repairs (preventive maintenance, abrupt repair, ex post facto repair)

Measures:

1. Definition or modification of inspection standards (areas, location, items, cycle of inspection)
2. Lubrication management, definition or modification of lubrication standards (cycle and replacement of oil)
3. Thoroughness in cleaning; thoroughness in initial adjustment → standardization
4. Definition or modification of spare parts management (order point, order quantity)
5. Improvement in the level of predictive techniques
   a) External inspection based on the 5 senses → measuring equipment (quantification)
   b) Standards of overhaul inspection (deterioration measurement)
6. Extension of parts life. Using point 1 above, and considering the unevenness in intervals and characteristics of breakdowns, estimate the life of parts and examine the manufacturers, mechanical structures (mechanisms), and materials used.
7. Thoroughness in corrective and preventive maintenance. Instead of simply restoring the equipment, add improved and modified elements to reduce failure frequency.
8. Changes or replacements of engineering drawings

a relatively high frequency of recurring failures, including significant repair time or production losses, or those failures whose repairs would constitute bottlenecks in the maintenance work schedule. Then analyze the steps involved in the appropriate maintenance work and investigate methods to reduce maintenance time. In addition, through the cooperation of the people directly involved in each department, create a list of machines that are poor in work efficiency for either operations or maintenance, and consider ways for possible improvement.

In addition, it is essential to review current organizational arrangements and task assignments of maintenance work, define educational and training requirements, and simplify the information flow routings, administrative procedures, and staffing situations.

One must identify problematic areas in work plans that are creating bottlenecks in overall work efficiency as well as formulate ways

TABLE 4.2 Measures to Improve Maintainability

Improvement of maintainability
(Reduction in repair time)

Key points:

1. Work activities consistently involving a large number of operational steps or high costs.

2. Work activities known to become bottlenecks in planned repair work.

3. Items that significantly affect product quality if there is a delay in repairs.

4. Work activities that require a high level of skill.

5. Work activities that are inefficient.

Measures:

1. Review sequence of overhaul inspection and maintenance work steps.
   Standardization...inspection checkpoints for operation steps, estimation time,
   and maintenance quality

2. Study work methods for reducing maintenance work time

   a) Block component replacement method (concurrent repairs)
   b) Availability and disciplined management of spare parts
   c) Availability of equipment needed (derricks, hoists, chain baskets,
      and transportation vehicles)
   d) Improvements in maintainability of the equipment (structures, layouts, installation
      positioning)--conducive to easy inspection, maintenance, lubrication, and cleaning
   e) Improvements in machine fixtures (ease, safety, speed)
   f) Improvements in the environment (heat, ventilation, dust, walkways, and lighting)

3. Improve managerial aspects of work efficiency

   a) Create adequate repair plans (networks)
   b) Availability of drawings and their accuracy (modifications or corrections to drawings, etc.)
   c) Improve human relationships among the work groups
   d) Clarify communication routes and hierarchy of positions for the work activities
   e) Availability of appropriate skill levels among workers and crews
   f) Education (multiskilled workers)
   g) Optimization of maintenance organization at the floor level

to improve maintenance plans. Investigate work done by outside repair contractors, and review the type of work that depends on contract repairs as well as the plausibility of the number of operational steps. Table 4.2 summarizes these measures for improvements.

4.4 PLANNED AND CHANCE MAINTENANCE

4.4.1 Importance of Planned Maintenance—A Behavioral Science Approach to Maintenance

1. FAILURE ANALYSIS SHOULD BEGIN WITH A STUDY OF MAINTENANCE ACTIVITIES

When we consult with companies about their maintenance departments, we often encounter work environments with an abundance of maintenance data and meticulous failure analyses, yet the number of failure occurrences remains the same. These are work environments in which the maintenance crew operates in a reactive mode to recurring equipment breakdowns, only managing to perform remedial emergency measures. In other cases, failures may occur at two locations simultaneously and not everyone needed for repair work can be summoned; the investigation of causal failure relationships takes too long; or too many abrupt breakdowns take place for enough time to be allotted to planned maintenance. When the reasons for these undesirable situations are considered, it is obvious that many things can be done to avoid them. Certain things can be done before breakdowns occur, there are cases in which simple changeovers or replacements can be done, and other problems can be avoided if care is taken during repairs. In thinking about the reason why a reactive, disorganized, "crisis" response to breakdowns tends to persist, we would suspect that breakdown analyses tend to focus lopsidedly on physical aspects, and they lack analysis of the relationships between failures and routine maintenance activities. Just as the definition of "failure" means that a given machine loses its specified function, the failure itself is a physical occurrence. Maintenance, on the other hand, means a human decision-making process and subsequent behavior, whose purpose is to ensure optimal machine conditions. The machine's reliability is assured through specific maintenance activities. Thus, the terms failure and maintenance must be clearly distinguished, and their causal relationship analyzed.

Figure 4.6 contains an analysis of maintenance and manufacturing department activities and failures. It shows the following:

- 12 percent of the failures occurred despite the fact that the maintenance department has a planned maintenance program;
- 14 percent of the failures occurred despite the fact that the production department has a self-initiated maintenance program in place, and
- Nearly 80 percent of the abrupt breakdowns are occurring in areas that have not been dealt with by either the maintenance or the production department.

Why then did such a high percentage of breakdowns occur? What can be done about them? If these breakdowns occurred despite ongoing maintenance activities, they require critical review and rectification. In this case, however, such study had not been carried out; hence the number of breakdowns is not surprising.

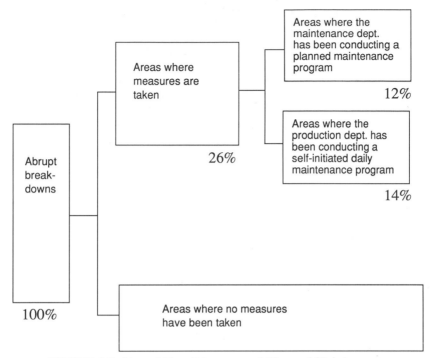

FIGURE 4.6 Abrupt Breakdowns and Planned Maintenance

2. UNLESS A MORE DETAILED MAINTENANCE PLAN IS MADE, NO RESULTS WILL BE ACHIEVED

An interesting picture emerges from a 1979 survey of equipment maintenance practices. The survey reports on companies with annual and long-range plans for promoting planned maintenance programs, comprising 83.3 percent of the total surveyed. The number of companies that have monthly plans comprises only 5.3 percent of the total. Another survey question revealed that more than one-half of the companies questioned are practicing total productive maintenance, but not even one-half of the total have plans for preventive maintenance. This means that for major maintenance activities companies have long-range plans, but that for all other kinds of activities their maintenance style is characterized by after-the-fact maintenance.

The annual maintenance plans, moreover, consist primarily of those focusing on securing the budget for their maintenance programs; the contents of the plans are consequently very brief. Each of the budget items seems to be intended mainly for maintenance activities that are difficult, expensive, or improvement-oriented. Such plans cannot be readily linked to remedial measures for breakdowns.

In the case of process industries, it should be sufficient to create a maintenance plan based on annual or long-range plans. In machine processing and assembly industries, however, there are just too many types of breakdowns. Detailed maintenance plans must be made, so that they can be connected to activities that reflect actual needs. Although the planning cycle may be shorter, preventive maintenance activities must be "live" and reflect real needs.

3. MAXIMUM ECONOMICAL MAINTENANCE FOR OPERATIONAL AND PRODUCTION DEPARTMENT NEEDS

A maintenance plan created solely by the maintenance department would tend to have a lopsided view of how maintenance should be conducted, since it would lack the perspective of the equipment users. Typical examples of this tendency are found in road construction plans for national highways and municipally maintained roads. Roads and highways are lot-parceled in terms of various types of work activities, such as electrical and water department construction work. Road construction work may be efficiently performed on an individual basis, but from the viewpoint of its users, it is often a nuisance. A maintenance method similar to this situation shall hereafter be called the "road construction method" maintenance.

In another example, the maintenance records of a machine's limiter switches show that each time a failure occurs in one of the switches, the maintenance crew scrambles to the area and tries to fix it. Despite the fact that there are only sixteen limiter switches in the machine, breakdowns and resultant production stoppages occur approximately once a month. Detailed study of the circumstances reveals that the life span of the parts involved is sufficiently long, but because of the scattered distribution of occurrences of the failures, they seem to be occurring at inordinately frequent intervals. In this case, a more effective maintenance plan would be one that is not merely reactive to failures, but one that reflects maintenance crew creativity as well. This may mean the implementation of a synchronized maintenance and tuning plan or a serialized parts replacement plan.

In the case of prioritized equipment (that is, equipment that has a higher maintenance priority), the cost of inspection and parts replacement should be less than the expense of losses incurred by parts failures. It should be noted that arbitrary increases in inspections and parts replacements do not necessarily lead to total failure elimination; they may instead merely incur costs and losses arising from work shutdowns.

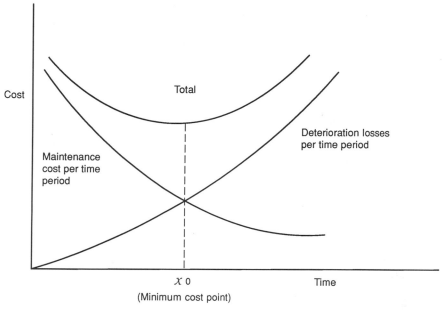

FIGURE 4.7 Optimized Parts Replacement Cycle

Depending on the types and conditions of operations, there are differing economical approaches to maintenance timing and methods. Various constraining conditions require an understanding of the most effective utilization of people, material objects, and money (see Figure 4.7).

4. THE MAINTENANCE PLAN DETERMINES THE QUALITY OF MAINTENANCE WORK

A singular aspect of maintenance activities is the difficulty of assessing the quality of work done. If maintenance work is poorly done, it may lead to a breakdown; because of the intervening time lag, however, it is hard to judge whether the breakdown was due to maintenance errors or defective parts. In other words, the quality of maintenance must assure the quality of the work itself. To accomplish this, each individual member of the maintenance crew must have a sense of responsibility and consider methods for preparing, executing, and validating his or her own work.

From the standpoint of efficiency, work planned for in advance can be accomplished far more quickly than work done in reaction to an abrupt failure. To raise the level of maintenance quality and efficiency, it is essential to create a maintenance plan and make tenacious preparations prior to its execution. When pursuing planned maintenance, the maintenance crew should say "let's curb break-downs." The benefits of creating the maintenance plan can be summarized as follows:

1. The number of operational steps can be identified and work can be routinized.
2. Human resource requirements can be planned so that required personnel are available.
3. Errors in the procurement of materials, spare parts, and subcontracting work can be prevented.
4. Quality can be checked and better materials can be procured.
5. By devising related work detail plans, schedules can be set so that they are coordinated with production plans.
6. Repair cycles can be identified so that measures can be taken in a timely fashion.
7. Standardized patterns for repair work can be identified, enabling the work to be done efficiently.
8. Simultaneous repair plans can be devised.
9. People's sense of responsibility can be encouraged.
10. Through planned work activities, a larger volume of work can be more efficiently accomplished.

4.4.2 What Is a Maintenance Calendar?

1. MAINTENANCE CALENDAR AND CHANCE MAINTENANCE

One of the excuses for stymied planned maintenance is "there is not enough time." The reason that there is not enough time is that the operations department would not stop the equipment just for the sake of the maintenance activities. Indeed, when the equipment is running fine and people are under the gun to meet the delivery date, the production activities should be given the first priority. Or should they? If an abrupt failure occurs, the equipment must come to a grinding halt and the net results would be the same as if they had stopped the equipment for maintenance. Moreover, in reality, there are many cases in which making a preemptive move and stopping the equipment for maintenance would prove to be much more valuable than otherwise.

This indicates that we must change our thinking with regard to our manufacturing equipment. As a rule, we keep a production plan chart. It is according to this chart that we keep our work days schedules and make plans for task-oriented activities, go about our production activities by allocating production equipment, people, and materials. This planning chart for production is the production (shop) calendar. If we look at it from another aspect, as we reviewed in a previous example, in order for us to accomplish our production goals, we must have the assurance that the production equipment is dependable. The plan for assuring this dependability is the maintenance plan. Since the production plan is shown in a calendar format, so too should this maintenance plan be shown in a similar fashion. The reason for this synchronism is that the plant operation and plant maintenance are two sides of a coin, that is, most of the equipment maintenance activities are carried out during the plant stoppage periods. We should make efforts to keep the maintenance calendar as complete as possible, since it reflects the maintenance plan that should be matched against the production plan. The relationship between the two is illustrated in Figure 4.8.

Another reason for not being able to find the time to do it is when the maintenance crew is totally swamped by abrupt breakdown calls, whereby they can hardly afford the time to get around to do the planned

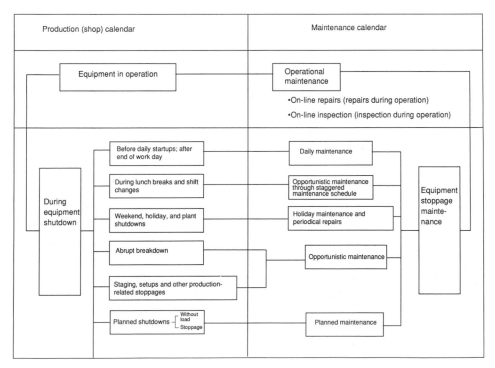

FIGURE 4.8

maintenance. In order to reduce abrupt breakdowns, we must stead-fastly pursue a thorough planned maintenance program and raise its level of standards. So long as people keep saying that they cannot even get started with such a program, no progress can be expected. To break up the vicious cycle, we need to have a time for equipment stoppage.

If we stop and examine if there really is no time for letting the equipment stop, we would realize that since the maintenance crew is going about the repairs for abrupt breakdowns, there are occasions when the equipment comes to a grinding halt. The point here is that we are not prepared to perform some other maintenance work while the abrupt breakdown situation is being answered. Once we start thinking along this line of reasoning, we would realize that there are opportunities for maintenance, such as during the times of product switchovers, setups, lunch breaks, and shift changes. It is not that we should allow the equipment to be stopped just for the sake of main-tenance work. Instead, what is important is that we have the positive attitudes toward pouncing on every opportunity that might arise.

2. PRODUCTION CALENDAR AND MAINTENANCE CALENDAR

The opportunities and the methods of carrying out maintenance activ-ities are described in Figure 4.8. Attempts are made therein to explain how they relate to the operations of the manufacturing equipment. One such opportunity is the operational maintenance (or on-stream maintenance) that is carried out while the equipment is in operation. From the engineering perspectives of PM, it is called on-stream repair (OSR) or on-stream inspection (OSI). It should be noted that there are many areas in the manufacturing equipment that can be repaired during their operation, moreover, there are many instances where in-spection and diagnosis can be carried out while the equipment is in operation.

The maintenance activities that are carried out while the equip-ment is stopped generally include the daily maintenance activities be-fore the work day begins and after it ends, or during nonworking days. It is regrettable that there is a tendency for the maintenance activities to concentrate on holidays and weekends. The problem stems from the fact that while the production crew stops the equipment during lunch break or shift changes, so too the maintenance crew would go out to have lunch and take breaks and so forth to the point of com-pletely foregoing the maintenance opportunities. The stoppages caused by abrupt breakdowns and setup times would create great opportuni-ties for maintenance. But in order to take advantage of these opportunities, it may often be necessary to plan the work schedule of the maintenance crew not by the production schedule calendar, but rather by a maintenance calendar. An example of this is a case in which the work schedule of the maintenance crew was changed to a staggered work schedule. In this case, the plant environment provided maintenance opportunities during lunch break and after the day's work. Although the stoppage time was between 1 and 3 hours, most of the maintenance work should fit into the time slot. It is also important in this case to make efforts in finely segmenting the maintenance work activities and also to prestage the maintenance work in a manner that would not interfere with the normal production activities.

With regard to another method, that is, planned stoppage, we are referring to a type of situation wherein the production department would not mind stopping the equipment even during regular work day hours. In general, the maintenance plan can be created within the maintenance department alone, however, it does not have the authority over equipment stoppages. Although that in itself often constitutes a reason for the planned maintenance to be stymied, after all, unless the production department would cooperate with a positive attitude by setting aside a TPM day or a maintenance day and willingly stop the equipment on its own, it would be a far cry from the planned maintenance to get accomplished. Unless the head of the production department makes this decision, the PM program cannot be advanced.

4.4.3 Methods to Promote Opportunistic Maintenance

Opportunistic maintenance means taking advantage of equipment stoppage times as they occur, in contrast to planned maintenance stoppage. Preparedness is the key to success; this method involves thorough investigation of such opportunities and their occurrences, focusing on the following:

1. When opportunities arise, which machines allow other concurrent repairs?
2. What are the precise opportunities, when do they arise, and how long do they last?

Maintenance activities should be arranged according to the scope of work activities (length of time needed for maintenance work). Records should be kept for each, as shown in Table 4.3.

TABLE 4.3

No	Table of records of opportunistic maintenance activities			Confirmation of Completion	Allowable time duration for opportunistic maintenance							Actual number of operation stops
	Issue date	Use of cards	Descriptions		Abrupt break-down	Planned mainte-nance	Margins needed for work details	Model change	FL Opening can of fluid	Removal of burrs	Others	
1	12/3	○	PR-1 Poor indication of LIA	○		●						2 x 4.0
2	12/8	○	Mechanical leakage in PR-2								○	
3	12/9		Malfunction of vacuum pump	●	●							1 x 2.0
4	12/10		No ringing of TEL	○							●	
30	12/18	○	Chain and cover connection of CRU	○		●						
31	12/19		Inspection of PSV			●						2 x 6.0
32	12/26		Poor indication of FT-113				●					1 x 0.5
33	12/29	○	Improvement in oil seal in the Pu section				○					

Month _____

Number of Occurrences: ( )

Number of Repairs: ( )

PDL	
PDR	
CPR	
QF	
Total	

4.4.4 Design and Highlights of the Maintenance Calendar

1. TYPES OF MAINTENANCE CALENDARS AND THEIR USE

Depending upon the characteristics and planning time perspective, maintenance calendars are divided into four types. Figure 4.9 shows a systematic method of using a maintenance calendar. In cases where detailed plans are necessary, such as in mechanical fabrication and assembly industries, a day per month or week should be regularly set aside for review of the maintenance plan by the maintenance and production departments.

If this crucial step is neglected, the maintenance plan could become nothing more than paperwork or a mere formality. Even if a maintenance plan is based on a simple system, it should be suited to

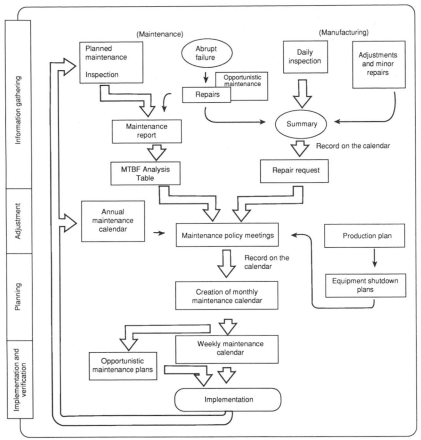

FIGURE 4.9 An Operational System for a Maintenance Calendar

the characteristics of the specific company, and its level of sophistication gradually raised.

There are four basic types of maintenance calendars, as follows:

1. Annual Maintenance Calendar
2. Monthly Maintenance Calendar
3. Weekly Maintenance Calendar
4. Individual Work Detail Plan

Each type is discussed in the following.

(1) Annual Maintenance Calendar

This plan covers the time frame that begins with the installation of a machine and ends with its shutdown or dismantling. It tends to focus on major maintenance work activities, such as overhaul, modification, and periodic maintenance type activities.

In this context, the budget allocation, vendor or contractor specifications, and major spare parts appropriations must be noted and meticulously planned. When people talk about a maintenance plan, in many cases, they are referring to this annual maintenance calendar.

(2) Monthly Maintenance Calendar

This calendar incorporates the annual plan and those plans with monthly planning cycles, such as plans for improvement and antibreakdown measures. It requires a balance between the amount of work activities and the number of maintenance crew members, as well as reliable followup on appropriation of resources. It is a calendar that requires the highest level of completeness.

(3) Weekly Maintenance Calendar

This calendar assigns the tasks that were planned in the monthly maintenance plan to individual crew members. It is used for staging work activities and for controlling their progress.

(4) Individual Task Plan

This plan applies to those situations, such as overhauls or modifications, that require longer time periods to complete. The task involved cannot be successfully managed unless schedules are made using a particular type of networking technique.

Types of Maintenance Methods

In the following, explanations are given for each type of maintenance method used in maintenance planning.

1. Periodic Fixed-Term Maintenance	After determining an optimum period of maintenance cycles, repairs and /or replacements are undertaken • For those machines in which time cycle periods can be easily determined with limited fluctuations.

	• For those machines in which periodic replacements without inspection have a significant advantage.
2. Predictive Maintenance	Inspections to investigate deteriorating conditions, and repairs based on such inspections • For those machines in which there is a distinct advantage in determining the maintenance period based on actual observation of the deterioration conditions. • For those machines that do not show any steady trend in deterioration, making it difficult to determine maintenance cycles. • For those machines that lack sufficient historical records, making it difficult to easily define their maintenance cycles.
3. Breakdown Maintenance	Repairs carried out after failure occurrences • For those machines in which this maintenance method offers an advantage, e.g., where the impact of breakdowns on other areas or losses are relatively minor. • For those machines with significant variation in deteriorations, thereby resistant to periodic inspections.
4. Corrective Maintenance	This refers to taking measures for extension of machines' life spans and reduction of their repair time and costs.

Maintenance methods can generally be classified under these four types.

Since periodic maintenance is carried out prior to actual failure occurrences, maintenance costs would tend to rise. However, in the case of breakdown maintenance, there are cases in which major breakdowns may prove less expensive if preemptive replacement methods are employed.

The breakdown maintenance method is advantageous for those machines that have a minor negative impact on other areas, or if measures are available that allow simple recoveries (such as shared pins, or nonfuse type circuit breakers). There is also a method that incorporates a redundant plan by setting aside spare machines to be used in the event of a breakdown.

Predictive maintenance forecasts the life expectancy of important parts and components through inspections, with the goal of maximum use for the longest possible time. It is a method that achieves the lowest maintenance cost and minimum losses from equipment breakdowns, and it requires development of techniques needed for predictive diagnostics.

Maintenance is a mixture of periodic and predictive maintenance activities. This distinction is often blurred when maintenance work entails repair work only, based on inspection results. It is also a method that is usually carried out in conjunction with the maintenance calendar; hence predictive maintenance is scheduled in accordance with the calendar.

2. CREATING AN ANNUAL MAINTENANCE PLAN

Since the annual maintenance plan is intended to ensure the reliability of equipment based on a long-range perspective, it usually extends beyond an annual scheme and should incorporate items that are maintained only once in several years.

The following are the steps in creating the plan:

1. Create a "shopping list" of items to be worked on
2. Select the items to be worked on by the annual plan—Prioritization
3. Trial estimation of maintenance periods—Reference use of MTBF Analysis Tables, etc.
4. Estimation of costs, work schedules, process steps, and maintenance time

5. Confirmation of tasks in which procurement is necessary—external contracts, purchased parts, certifications, etc.
6. Review of maintenance cycle periods
7. Creation of tabulated plans and distributions

The first step is the key: creation of a list of items that need to be maintained. Identification of work items for annual maintenance is the most important task—a list that is seemingly without limit, but one that should be reviewed at least once a year.

Method of Identifying Items that Need Maintenance

1. Legal regulations—Those items that pose risks in terms of safety and accident prevention.
2. Owner's manuals—Those items about which maintenance can be anticipated from the time they were designed. One should be aware that some equipment may have undergone changes since it was designed, or some may have been used in excess of what its designers might have anticipated. Hence these items should be given special considerations.
3. Results of precision maintenance or deterioration assessment tests—These are determined on the basis of the target values of equipment precision maintenance levels.
4. Breakdown records—The maintenance items needed to prevent breakdown recurrences.
5. Last year's annual maintenance plan—Those items that underwent changes in the maintenance cycle periods.
6. Opinions of Site Personnel—Due to abnormal equipment conditions, a need is perceived for taking some sort of measures.

Figure 4.10 presents an example of an annual maintenance plan table.

3. CREATING A MONTHLY MAINTENANCE PLAN

A monthly maintenance plan interprets the annual maintenance plan in terms of a monthly time frame. It is a specific plan of execution indicating the particular goals of the annual plan to be pursued under a set of given constraints. The following are the procedural steps for creating a monthly maintenance plan:

Equipment: Continuous copper plating (Machine mode no.) Cu-9

Annual Maintenance Planning Table (Maintenance Calendar)

/ denotes plan
x denotes confirmation check of accomplishment
Created: 3/29/83

Equipment unit	Location	Maintenance Contents	Maintenance period	No. of maintenance process steps (man-hrs.)	Parts used	Costs	Assigned to:	1	2	3	4	5	6	7	8	9	10	11	12	Remarks (book record to be used for work activities, etc.)
Pump	Circulation pump for the elecrolytic solution	Overhaul maintenance	1/ 6 mos.	2 people x 8 hrs.			Maint.		/						/					
	Other pump	"	1/yr.	2 people x 100 hrs.			"								/					20 units 5 hrs./unit
Valve		Replacement of sealing gasket	1/yr.	4 people x 40 hrs.	* Sealing gasket	$1,000	Maint.								/					30 locations
		Tightening	1/ 4 mos.	2 people x 15 hrs.			Maint./ Prd.				/				/					"
Plumbing	Water pipes	Inspection (cleaning)	1/ 3 yrs.	2 people x 4 hrs.			Maint.								/					Previous maintenance carried out: 8/1987 Next maintenance planned: 8/1990
Electric supply roll	No. 1	Replacement of roll(s)	1/ 5 wks.	3 people x 3 hrs.	Roll	Polishing Correction $400	Maint.					/								Monthly calendar
		Replacement of bearing(s)	1/ 6 mos.	3 people x 3 hrs.	* Pillow blocks 2 units	$67	"											/		Roll-polishing is contracted out (monthly calendar)

Note: Inspection to be conducted by means of the inspection checksheet.

Note: Maint. = maintenance dept.; Prd. = production dept.
Note: August month cycle has a periodical repair/maintenance.

FIGURE 4.10

Creation Steps

1. Organization of items to be maintained
2. Estimation of process steps, maintenance time, and costs
3. Prioritization—degree of urgency, legal regulations, date specified for actual work, etc.
4. Organization of constraining factors
5. Routinization of work load and specification of work schedule
6. Appropriation

---

(1) Organization of Items to Be Maintained

The organization of items to be maintained for the monthly maintenance plan should be carried out as follows:

1. Carry over the tasks for the next month from the annual maintenance plan.
2. Determinations should be made based on analysis of the maintenance and breakdown repair records as well as the inspections.
3. Determinations should be made based on daily inspections of the production department and results of minor repairs.
4. Plan for alteration of equipment layouts and design of machine fixtures, etc.
5. Plan for improvements of quality and safety.

The most important stage in maintenance occurs during selection of the areas and items for which plans are to be made. It is necessary, therefore, to hold a regular monthly meeting with all the people involved in related areas of concern. These meetings can also provide the opportunity to distribute tasks that can be accomplished by individual departments. It often happens that items requested by another department or section, or the results of failure analysis, lack specific details. In these cases, the maintenance department should decide on the measures to be taken.

When detailed and specific monthly or weekly plans are to be made, maintenance records must always be used as reference. If this is done, the written quality of the records improves, a clearer understanding is obtained of how the records should be used, the overall quality of the records rises, and as a result, the maintenance plan itself improves. It is for these purposes that the MTBF Analysis Table was devised.

When the maintenance plan is analyzed, the review should be based on maintenance records and consideration of failure characteristics, as well as the following:

- What are the crucial parts or components?
- What kind of failure modes can be anticipated?
- How long is the maintenance period?
- How long does it take for maintenance, what is the degree of difficulty involved, and what tools and personnel skills are needed?
- Are spare parts available?
- What is the key point of maintenance quality?
- Is there a need for improvement in maintenance for this specific case?

It should be noted that the production department usually requests items that are related to actual problems that occur. It is necessary, therefore, for the maintenance department to interpret these requests and devise the specific measures to be taken.

If monthly maintenance planning tasks are followed as just outlined, many relevant and high-quality maintenance plans should be readily devised that are conducive to taking detailed preemptive actions.

(2) Estimation of the Process Steps, Maintenance Time, and Costs

When the specific content of a maintenance plan is determined, then the number of steps involved in the plan and the cost of each step can be estimated. While personnel are learning this estimation process, absolute quantitative accuracy is less important than the learning process itself. Maintenance leaders should continually sharpen their skills to achieve more accurate estimation levels. The number of process steps involved in a maintenance task may deviate by up to ± 15 percent. Unless a general standard has been set, the estimation should be made by an expert. If the estimation is difficult even for an expert, then instead of aiming for a single-point estimation technique, a 3–point estimation technique should be used.

(3-point time estimate) = [(Pessimistic) + 4 × (Average) + (Optimistic)] ÷ 6

(3) Organization of Constraining Factors

Constraining factors can be defined as follows:

1. Factors determined by the production department, such as planned equipment shutdowns, specific predetermined maintenance days, etc.
2. The available number of specially skilled workers in the maintenance crew and establishment of a delimiting frame for emergency failure repairs.
3. Targeted maintenance costs.
4. Personnel skill levels required.
5. Whether or not the maintenance can be carried out while the equipment is in operation, or only during holidays and weekends.
6. Whether maintenance must be contracted out, or whether it can be done in-house.

(4) Appropriation

Monthly plans most commonly fall behind schedule because of problems related appropriation. This problem typically takes the form of forgotten or delayed initiation of maintenance, insufficient followup, and so on. Procured parts should be inspected for quality upon receipt, and it should be known where they are stored.

Subcontracting reservations should be made well in advance. Meticulous attention should be paid to various details: not neglecting the initiation of appropriation steps for equipment and tools to be used for maintenance, or communication with vendors regarding plan alterations, cancellations, or postponements. With regard to maintenance that can only be carried out when the machines are shut down, this work must be closely coordinated with the production department, especially when it is trying to meet preclosing production quota that may necessitate changes in the plans.

Key points in actual use—maintenance plan with flexibility. The maintenance department is a "rescue squad" or "fire company" for the production department, and readiness and responsiveness in the event of emergency breakdowns are absolutely necessary conditions for its operation. But even more important are the maintenance department activities designed to prevent accidents. Thus the maintenance department must be able to respond to emergency situations and still be able to pursue a planned maintenance program. Achieving these goals requires ingenuity and creativity.

Point (1): Divide maintenance activities into three types.

1. Those maintenance activities that are performed while the machines are stopped and that take a long time.
2. Those that are performed while the machines are stopped but that can be accomplished within a shorter period of time (opportunistic maintenance).
3. Those activities that can be performed while the machines are running.

The activities in type 1 require a machine shutdown plan; such activities necessitate the highest priority management, since they call for a significant amount of preparatory work in order to be accomplished flawlessly. This preparation includes adjustments to changes in the production plan, subcontracting procurements, and parts appropriations. Thus it is important to divide the tasks and the external liaison, as much as possible. Each subdivided task, moreover, should be simplified so that anyone can step in and take over as needed.

The work activities in type 2 aim to take full advantage of the time when the machines are stopped. To this end, tasks that can be accomplished within one to two hours should be listed and prepared for. When an abrupt breakdown occurs or a lunch break is taken, or during any stoppage time that might occur during an ordinary day's work, that time can be used in an aggressive manner.

The work activities in type 3 are inspections, parts fabrication, or external appropriation type activities. These activities should be planned for, and the maintenance crew's spare time should be used to implement the maintenance plan efficiently.

Point (2): Create a plan that allows easy schedule alterations.

There are enterprises in which abrupt breakdowns occur often, and the maintenance plan quickly becomes "punctured" and dies. In these cases, people tend to reduce the maintenance plan to focus only on a few vital items, or divide the maintenance crew into a plan execution group and an emergency "scramble" group. To reduce the amount of abrupt breakdowns, the number of items to be planned is usually expected to increase. This practice, however, is not necessarily a desirable approach. Cases are also seen where the plan is written up on a sheet of paper. This might serve its purpose to a certain degree, but only as long as the number of planned items is small. If the number is large, it is difficult to make changes.

It should be kept in mind that routine changes will occur in any maintenance plan. An abrupt breakdown, therefore, can be taken as

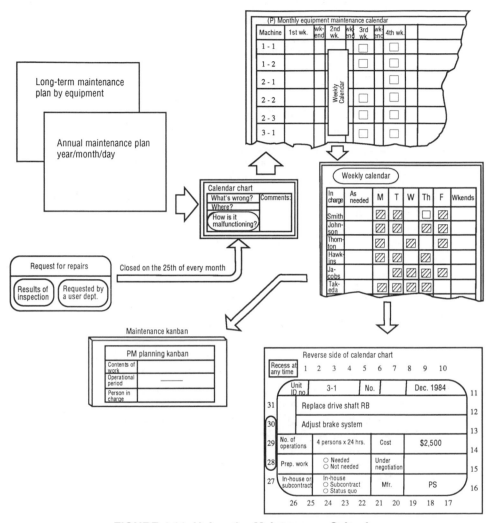

FIGURE 4.11 Using the Maintenance Calendar

an opportunity, and the plan is then modified accordingly and in an aggressive manner. To facilitate this possibility, the plan should be kept both consistent and responsive to change. For example, maintenance tasks should be recorded on a card, making it possible to conduct management by observation (see Figure 4.11).

Point (3): To pursue planned maintenance efficiently, the following important areas of the maintenance system itself should be considered:

• Staggered work hours and opportunistic maintenance;
• Task assignments among maintenance crew members;
• Integration of various maintenance jobs.

The goal of the staggered work hour system is to take full advantage of any time when the machines are stopped. For machines with a high work load, however, it is often true that not much can be practically done unless lunch breaks or overtime work hours are used for maintenance work. To cope with such a situation, tasks can be accomplished by staggering the work hours of the maintenance crew.

In plants where abrupt breakdowns occur frequently, it is often found that maintenance crew members are not familiar with the specific details of the machines. To deal with this situation, the maintenance crew can be assigned responsibility for certain plant machines or locations. This would improve communication between the production and maintenance units, leading to cooperation in devising measures to be taken against breakdowns. Both groups would become well versed and knowledgeable about the assigned equipment or locations. This method of assigning clearly defined and ongoing responsibility can be effective in promoting maintenance plans.

Some repairs cannot be undertaken unless a combination of certain jobs or skills is available. In enterprises where job classifications are stringently segregated—such as electricians, mechanics, plumbers, etc.—there is a tendency toward task fragmentation, resulting in lower total effectiveness. For such cases, an integrated maintenance department should be designed, by having groups of personnel with different skills and specialties work side by side, or by organizing combined groups. The resulting benefits could include more effective maintenance plan preparation, a higher rate of skill acquisition deriving from joint reviews and actions, and a reduction in mutual complaints and recriminations.

Management of Progress and Evaluation
The methods of using the monthly and weekly maintenance calendars are shown in Figure 4.11. The weekly maintenance calendar defines specific work details in terms of operational steps. Even if maintenance work is to be done after plans are made, problems may arise and the planned schedule could go awry. Therefore, thorough understanding of the circumstances surrounding each specific case is needed, as well as the courage to respond dynamically to ongoing changes.

A maintenance plan should be evaluated in terms of the following:

1. The degree of the plan's contribution to the operational departments: Evaluation can be based on answers to questions of whether

the equipment's reliability level has risen; whether the number of unexpected failures has decreased; or whether the level of availability of the equipment or the MTBF has improved.
2. The extent of efforts made by the maintenance department: Evaluation can be based on how many maintenance plans have been prepared; how much they have accomplished; how much improvement there has been in time estimates; and how much the MTTR has been reduced.

It is also important to consider total maintenance costs. A number of evaluation methods exist, e.g., the DuPont method. If the evaluation method is too complex, however, people can lose sight of the objectives. Thus, simpler cost evaluation methods are better.

4. DEVELOPING PLANS FOR "ISOLATED" WORK

Isolated work means any kind of work that takes place while a certain machine is stopped for a specified period of time, i.e., work done during an idle period, during a periodic maintenance period, or during a shutdown. Because such machine stoppage may mean a significant drop in the overall production level, a carefully detailed and efficient plan is needed.

(1) Characteristics of Planning for Isolated Work

1. The most important task is to reduce the length of the work period.
2. A significant cost is incurred.
3. Plans should be made for procurement of materials, human resources, and subcontracting.
4. The overall work is comprised of many smaller work units.
5. A careful quality check of the work should be maintained.
6. Depending on the level of deterioration of the machine in question, the nature of the work may change.

(2) Making an Isolated Work Plan

A detailed plan should be made based on the results of previous inspections and modifications, and the characteristics of the isolated work, as just itemized. For efficiency, a step-by-step operational procedure plan should be made of all the work details. For manageability, a detailed work schedule should be drawn up, including specific dates for completion and delivery of each step and operation. The Program Evaluation and Review Technique (PERT), which is often used for scheduling, is illustrated in Figure 4.12. This technique can

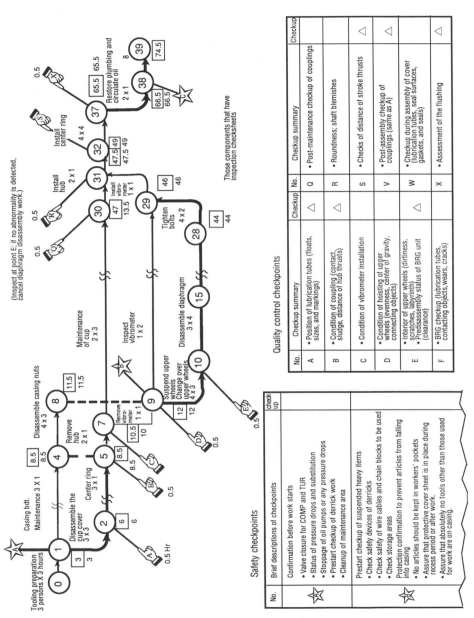

FIGURE 4.12 A Network of Work Details for Periodic Maintenance of a Compressor

188

highlight which group(s) of work-related activities need management to meet their delivery deadlines and can help define the key factors in their quality and safety. PERT enables graphic and comprehensive management of all these factors, and helps ensure needed startup of specific work details.

The basic ideas in creating a schedule are:

1. To create a master schedule on which detailed schedules are based, to ensure on-time delivery.
2. To synchronize the speed of work progress among the different departments involved.
3. To standardize and routinize the amount of work to be accomplished daily.
4. To reduce the time needed for work to be done.

(3) Undertaking the Isolated Work

Since support of isolated work requires the coordinated efforts of the maintenance, operation, engineering, and procurement departments, each department should appoint someone to take charge and maintain coordination through regular meetings. Everyone should also be ready to make dynamic changes in the nature and content of the work as the inspection results are evaluated.

4.5 DIAGNOSTIC TECHNIQUES AND PREDICTIVE MAINTENANCE

4.5.1 Values of Equipment Inspection Activities and Current Issues

The activities involved in equipment inspections have been emphasized as the foundation of PM activities. We shall now reassess their values and discuss future directions from the standpoint of "preventive maintenance."

1. PURPOSE AND VALUE OF EQUIPMENT INSPECTIONS

Equipment inspections are indispensable functions of PM activities. Predictive maintenance for high-priority equipment must include periodic inspection of the equipment in order to detect and take appropriate measures against any deleterious conditions that may lead to abrupt failures or deterioration in product functioning and/or quality.

It is crucial that the sources of problems be removed, adjustments made, or recovery measures taken while the problems are still in an early stage. The goal of predictive maintenance, therefore, lies in economical measures taken against equipment deterioration. A brief summary of these activities is shown in Figure 4.13, which identifies three

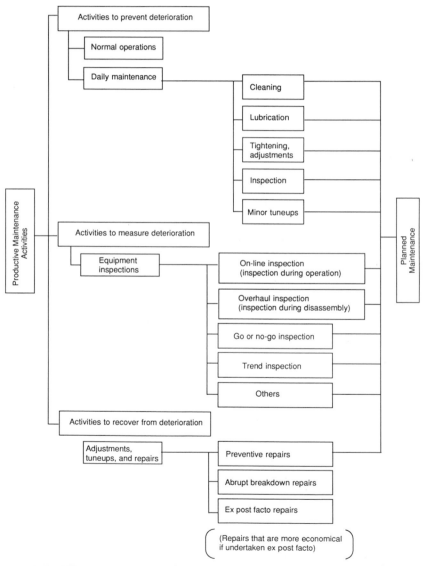

FIGURE 4.13 Measures Taken Against Equipment Deterioration

main types of measures to prevent equipment deterioration, listed in order of importance:

1. Activities to prevent equipment deterioration (daily maintenance, normal operations);
2. Activities to measure equipment deterioration (predictive techniques);
3. Activities to recover from equipment deterioration (repairs, adjustments).

Two kinds of situations are known to affect significantly the extension of equipment life span: (a) operators engaged in production activities are trained to understand the equipment's functions and structures, and (b) through daily maintenance, the equipment is lubricated, cleaned, and given necessary minor adjustments and tuneups.

Apart from inspection for daily routine maintenance, measurement of deterioration can be considered to be a type of activity that assesses abnormal conditions and deterioration using advanced inspection techniques to predict the remaining useful life span of the equipment. The following are some examples of equipment inspections during operation: functional deterioration may be assessed on the basis of vibration characteristics; shocks received by the receptor bearings of an axle shaft may be measured by means of a pulse tester; or stress factors may be measured for given parts or components so as to investigate the appropriate measures to be taken. Overhaul inspection tasks that require equipment stoppages include alignment measurement or inspections using nondestructive testing equipment, such as magnetic blemish detectors, a depthometer to measure the extent of "turtle" cracks, ultrasonic thickness gauges, tubular blemish detectors, and so on.

The purpose of these measures is to achieve a rational and efficient planned repair program (preventive repair) aimed at economically preserving the equipment from deterioration. A planned maintenance program would reduce the proportional costs of production losses from stoppages caused by abrupt equipment failures and/or losses due to industrial accidents and repairs. Equipment inspection activities are designed mainly to raise maintenance efficiency by taking appropriate steps to prevent deterioration, based on the results of such inspections. Thus, unless remedial measures or adjustment methods are effective, accurate detection of problems alone will not be sufficient for reducing abrupt breakdowns or the number of repair operation steps. In sum, inspection activities are necessary; however, their value is measured only in relation to the methods used for repairs, adjustments, and tuneups.

2. FAILURE MECHANISMS AND FAILURE MODES

Why do breakdowns occur? Just as daily life involves people in a struggle against mental and physical stresses, so, too, in the mechanical world, materials, parts, and products are under stress, whether in use or left idle, and are exposed to hazards of deterioration and breakdown. These stresses are of two main types: operational and environmental. Operational stresses (or functionally derived stresses) are those stresses that are the unavoidable result of the operation or functional use of some products (i.e., mechanical stresses exerted on internal combustion engines, or electric power on electric bulbs). Environmental stresses, on the other hand, are those exerted by the external environment (such as the ambient temperature, humidity, surrounding vibrations, external impacts, etc.).

Breakdowns occur when the extent of the stresses exceeds the strength of the products or equipment (see Figure 4.14).

To prevent breakdowns, there must be a margin between the stresses on an object and the endurance strength of the object, that is, there must be a safety margin defined and set aside for the equip-

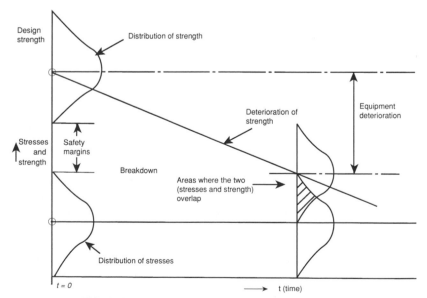

FIGURE 4.14 A Model of Stresses and Strength

TABLE 4.4 Failure Modes, Mechanisms, and Stresses of an H-101 Tube

Stress			Breakdown mechanism	Failure modes
Operational stress	Environmental stress	Time requirements		
Conduction, flow, radiation	Gradual stresses	Cumulative operation hours	Gradual stresses	Cracks
Energy-added fabrication	Destruction (by temperature, pressure, heat, fluid, etc.)	Repetitions (temperature rises, temperature drops, speed in changes and frequency of changes)	Damages • wear in gauge thickness • ellipsoidal warpage	Breaks
Structure (diameter, thickness, length, etc.)			• macroorganismic changes	Warping
Material	Change in temperature, pressure, etc. (cycle shock)		• microorganismic changes • compositional changes	Heat exposure
Thermal expansion		Cumulative load	• lowered thermal-resiliency level against high and normal temperatures	Wetness exposure
Slackening	Composition of processing gases, and their changes	Cumulative hours of use of catalysts and load		Corrosion
Catalyst (property)		Circumstances of temporal changes in the environment	• lowered hardness level	Wear
Deterioration during transportation and storage	Composition of flue gases and their changes		• lowered strength level against gradual ruptures	Congestion
Mechanical load	Gaseous distribution	Duty cycles	Corrosion	
	Operational load and its changes	Temporal deterioration	Wear	
			Fatigue	
			Excess heat exposure	
			Excess burden load	
			Degraded catalyst (palletization, etc.)	

(Notes) Depending on the environmental and operational conditions, breakdowns are considered to occur when the cumulative amount of energy exceeds a certain threshold, or when deterioration progresses beyond a certain threshold. The factors inducing these circumstances are generally called stresses. (It should be noted that, in mechanical engineering, the term "stress" refers to applied dynamics and kinetics. Here, however, it is used in a broader sense.)

ment, product, or part. But if the margin is too great, the object may be designed for durability in excess of what is required, resulting in higher costs and hence unsatisfactory function.

The physical, chemical, and mechanical processes that contribute to breakdowns are called the mechanism of breakdowns, and the failure modes are the phenomena that emerge as a consequence of the breakdown.

Analogous to illness in human beings, the failure mechanism corresponds to the pathology and the failure modes to the symptoms.

Maintenance technicians should always be involved in the investigation and analysis of failure mechanisms as described before, and in efforts to reveal the failure modes. This means developing diagnostic techniques, i.e., techniques to detect deterioration. In other words, diagnostic techniques refer to techniques for quantifying the stresses on machines, their deterioration, and their breakdowns, so as to predict and anticipate causes and consequences (see the examples shown in Table 4.4).

Depending on the environmental and operational conditions, breakdowns are considered to occur when the cumulative amount of energy exceeds a certain threshold, or when deterioration progresses beyond a certain threshold. The factors inducing these circumstances are generally called stresses. (It should be noted that, in mechanical engineering, the term "stress" refers to applied dynamics and kinetics. Here, however, it is used in a broader sense.)

3. CHARACTERISTICS OF IN-OPERATION (ON-LINE) INSPECTION, ANALYSIS, AND INVESTIGATION FOR SYMPTOMS OF ABNORMALITIES

Two types of equipment inspections have been discussed previously: one to be done while the equipment is in operation and the other while it is shut down. Not only does disassembly-type inspection require a considerable number of operational steps, but the determination of the inspection cycle poses quite a complex theoretical problem.

Regardless of how difficult a task it may be, inspection cycles must be based on the opinions of the engineers and of equipment manufacturers, as well as on past experience, in order to prevent abrupt breakdowns (in contrast to breakdowns caused by functional and/or qualitative deterioration).

However, if some symptoms can be detected during equipment operation (instead of repeating certain inspection procedures at predetermined intervals), and if remedial measures can be taken in the optimized repair cycle (instead of during inspection intervals), significant results can be expected. This is one reason why so much is at stake in the development of inspection techniques while the equipment is in operation.

In-operation inspection is a technique for estimating the status of internal elements based on observations of external phenomena in a given piece of equipment. In this sense, it is distinct from both disassembly inspection techniques and techniques used to analyze the cause of breakdowns. The in-operation inspection technique is similar to that used by doctors who diagnose problems by means of stethoscopes, x-rays, hematological tests, electrocardiograms, urine tests, and so on. In other words, external diagnostics is based on finding the causal relationship through the results of analysis.

In many cases, a machine may show symptoms before a full-fledged breakdown occurs—giving a sign that prompts observers to comment on the "strange" or "weird" behavior of the operation. Specialized analytical study can allow the early recognition of this type of symptom, enabling a breakdown to be predicted. To prevent a breakdown

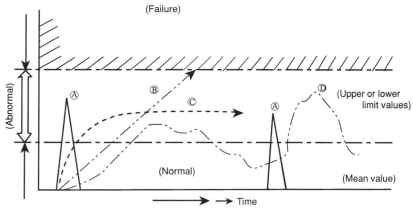

Types of symptoms

Ⓐ Abrupt type (Example: Lodging of a foreign object in valves and dislodgement)

Ⓑ Progressive type (Example: Leakage through a pinhole or flange)

Ⓒ Stable type (Example: Leakage through a pinhole or flange)

Ⓓ Frequent type (Example: Repeating or reciprocating types)

FIGURE 4.15 A Conceptual Model of Abnormal Symptoms

within a huge equipment complex, the operators—who are in continuous contact with the equipment—must carefully observe and understand the symptom(s), thoroughly and professionally analyze the cause of the aberration in cooperation with the maintenance technicians, and be able to predict an inevitable breakdown based on identification of the relationship between the symptoms and their causes (see Figure 4.15).

4. PROGRESS IN EQUIPMENT INSPECTION ACTIVITIES ACCOMPANYING EVOLUTION OF PM

A considerable amount of time has passed since the PM concept was introduced into the industrial world of post-war Japan. When examining a recent survey of PM activities in companies long considered leaders in PM, however, we observed that diagnostic techniques, which should be the technological foundation of PM, have seldom been developed—with only a few exceptions—by the maintenance engineers or technicians.

At present, there has been no change in the main concept of inspection, namely, inspection associated with a floor patrol "beat" as the centerpiece of daily maintenance. (We should, however, recognize the value in this concept.) With regard to the patrol inspection method, analysis of the causal relationship between external phenomena and internal mechanical elements is carried out according to the inspection crew's skills, and the whole process is entirely at the mercy of individual skill levels. There has been little standardization in this area.

When we ask questions about equipment inspections overseas, people usually tend to describe the activities of what they call "inspection engineers." In many cases, these activities include predicting the extent of equipment deterioration based on vibrometer measurement; examples of maintenance improvements using strength measurement meters; optimized predictions of repair cycles of the drive axle receptor bearings based on shock pulse testers; or development of various diagnostic techniques and improvement maintenance activities. In Japan, all these tasks are done by maintenance technicians.

For some time, company practice in Japan included hiring able graduates from technical high schools and sending them to the "front lines" of maintenance action as equipment "physicians." These young people had a strong sense of dedication and commitment, and initiated activities to conduct planned equipment maintenance. Despite their determination and efforts, however, the techniques needed to predict the extent of deterioration proved beyond their abilities. Such predictions called for extensive experiential know-how, and without the needed trust and support of the production and repair departments, the efficacy of the young peoples' activities was questioned. Ultimately, in many of these cases, the functions of this group diminished and eventually disappeared—even though these companies were faced with a shortage in human resources against a backdrop of successive facility expansions.

Today, maintenance technicians are required to have the ability to improve equipment maintenance as well as production. Beyond this, the development of diagnostic techniques, together with the wider use of instrument-dependent inspection techniques, has come increasingly to form the basis of economical maintenance plans. In the future, these needs are expected to increase even more. Figure 4.16 shows a systematic view of diagnostic techniques. These diagnostic techniques should be utilized at each individual phase within the equipment's life cycle.

The instruments and tools used in these diagnostic techniques are not in high demand in the marketplace. Although manufacturers fully understand the industrial community's need for these instruments, the tools' design requirements tend to be too specific, and the market for the instruments is too fragmented to justify investing in the sizable

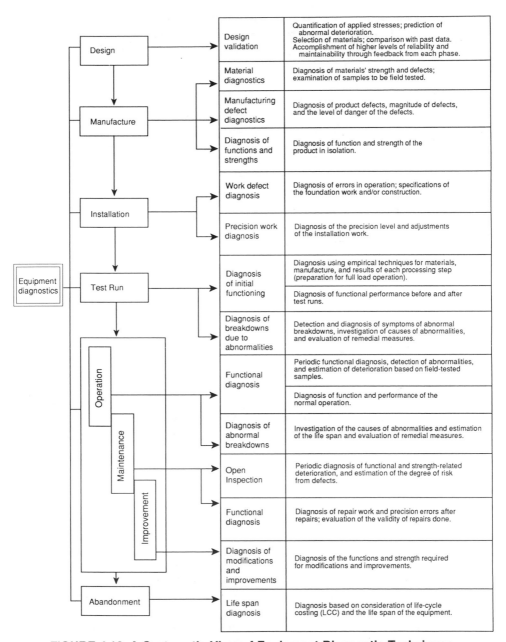

FIGURE 4.16 A Systematic View of Equipment Diagnostic Techniques

197

development costs required. Consequently, we have to make do with "off-the-shelf" general equipment and instruments. This being so, maintenance engineers face a strong need to develop unique diagnostic techniques suited to their own plant equipment. This is one area in which it is difficult to work with technologies purchased from outside sources. The diagnostic techniques developed by technicians of individual companies have recently begun to come to public and general view. Henceforth, activities to improve the rate of equipment effectiveness mean competition among different diagnostic techniques.

4.5.2 Predictive Maintenance

1. CONCEPT OF PREDICTIVE MAINTENANCE

The concept of predictive maintenance referred to here is based on the value of equipment maintenance aspects of PM activities in Japan and the development of inspection technologies. It is a philosophy of averting the tendency toward overmaintenance (i.e., excesses in maintenance and repairs) to which conventional approaches to predictive maintenance are prone. It is also a philosophy of promoting economical PM activities based primarily on an engineering investigation into optimized maintenance cycles. Although the term "preventive maintenance" may sound relatively new, in reality, it has been used in Japanese industries for quite some time. The essence of PM activities lies in equipment inspection technologies and therefore focuses on efforts to develop them.

As we investigate how planned maintenance is actually carried out using the conventional predictive maintenance method, we realize that there are just too many problems in determining the maintenance periods. To be more specific, some illustrations are in order.

1. The number of planned maintenance operations actually carried out is significant, regardless of the fact that because of the occurrence of abrupt breakdowns, neither the statistics used as the breakdown occurrence frequency nor the actual contents of the samples are sufficient.
2. There are frequent cases in which diagnoses of internal conditions are postulated despite the fact that they are based primarily on the external inspections using the five senses, and after the suspected facilities are overhauled, no significant problems are found within them.

3. Insufficient reviews are made when inspection cycle periods are to be determined. People seem to be resigned to the idea that there is no better method than simply periodic maintenance. For example, methods of determining the maintenance cycle periods may require it to be done every three months, every half year, one year, two years, three years, etc.

In complex production operations some key questions about planned equipment maintenance should be considered on an individual basis. Given the differences among individual plants in production quantity, scope of losses arising from work stoppages, and labor conditions, which equipment should be maintained according to a plan? Which should be removed from the list of planned maintenance items? Does the presence or absence of a clear determining factor for maintenance periods make a difference in deciding whether the equipment should or should not be maintained via a plan? The ambiguities that exist vis-à-vis determination of maintenance periods have greatly affected the rate of decline of equipment utilization and increased maintenance cost.

The predictive maintenance method does not entail complex and difficult concepts. Predictive maintenance simply means sensing, measuring, or monitoring any physical changes in facilities, predicting and anticipating failures, and taking appropriate remedial actions.

2. GENERAL TECHNIQUES FOR PREDICTIVE MAINTENANCE

To improve the rate of equipment effectiveness and reduce maintenance costs, it is necessary to promote corrective maintenance, to conduct thorough investigations into predictive maintenance, to carry out breakdown maintenance based on sound economic justifications, and to promote preventive maintenance.

In Japanese PM activities, we also need to emphasize the importance of predictive maintenance based on equipment inspection and examination techniques. The role of maintenance technicians should be to improve the rate of equipment effectiveness and to reduce maintenance costs through inspection techniques that would be unrivaled in their level of competitiveness.

3. GOALS OF PREDICTIVE MAINTENANCE ACTIVITIES

The following are the goals of predictive maintenance:

1. To determine the optimal time period for maintenance;
2. To reduce the amount of work for preventive maintenance;

3. To prevent abrupt breakdowns and reduce unplanned, ad hoc maintenance work;
4. To extend the life span of machines; their parts and components;
5. To improve the rate of effective operation of equipment;
6. To reduce maintenance costs;
7. To improve product quality;
8. To improve the precision level of equipment maintenance.

4. TECHNICAL TOPICS OF PREDICTIVE MAINTENANCE ACTIVITIES

1. Study of vibration
2. Study of generated temperature
3. Study of abnormal pressures and stresses
4. Study of wear and deterioration
5. Study of alignment
6. Study of corrosion and erosion

These are to be used as topics of investigation. The predictive maintenance philosophy can be promoted by focusing on the functions and structures for all the prioritized equipment in light of those topics. If the off-the-shelf standard measuring equipment available in the marketplace is not suited to the methods required for detection, measurement, and monitoring of the equipment, then the maintenance technicians should take on the challenge of developing the needed instruments or equipment.

4.5.3 Examples of Promoting the Predictive Maintenance Philosophy

To promote an effective predictive maintenance program based on the use of diagnostic techniques, a systematic approach is necessary to avoid haphazard activity.

Past breakdowns must be analyzed and categorized according to the topics of investigation listed before. Figure 4.17 illustrates the causes of problems in a chemical processing plant and the major breakdowns of a semiproduct manufacturing industry, respectively. It was found that, in all cases, diagnostic techniques were being used in one way or another. Thus, we must focus on some specific components of the equipment that were identified in the investigative process, enter them in the maintenance calendar, and proceed efficiently with planned maintenance. Figures 4.18 and 4.19 give examples of organized ways

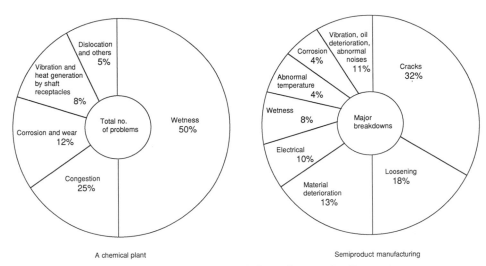

A chemical plant   Semiproduct manufacturing

FIGURE 4.17 Examples of Failure Causes by Category

to map the relationships among the components to be diagnosed, their symptoms, and remedial procedures.

4.6 IMPROVEMENT OF MAINTAINABILITY

Maintainability means ease of maintenance. Improving maintainability, therefore, refers to activities that can reduce the amount of time used for maintenance. These activities comprise one of the two main pillars of PM (i.e., improvement in equipment reliability and improvement in equipment maintainability). Thus, it is a cardinal theme, particularly for the maintenance department.

4.6.1 Recognizing the Characteristics of Maintenance Work

Activities such as daily maintenance, inspection, and maintenance repairs tend to have a low level of work efficiency as compared to work activities that are directly related to production. An educated guess of the difference between maintenance activities and the main production activities would estimate the actual number of operational steps at about 50 percent fewer for the latter. This is because the number of staging, setup, and other residual steps, in addition to those asso-

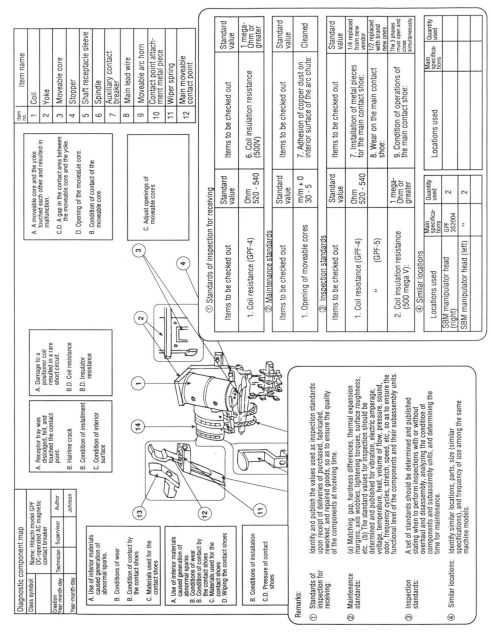

FIGURE 4.18

202

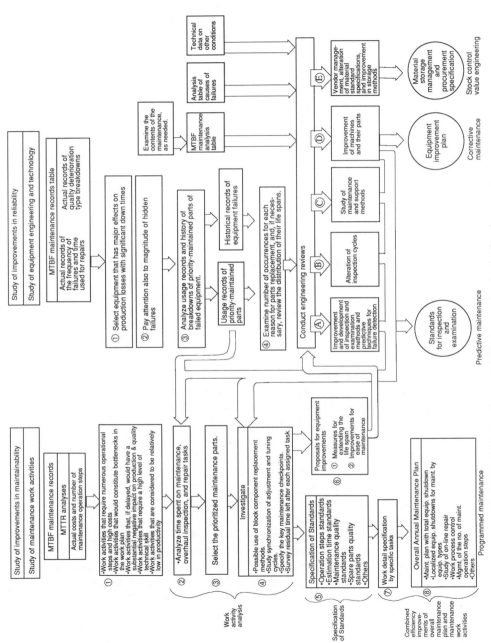

FIGURE 4.19 Equipment Efficiency Analysis Deployment

203

ciated with the usual production cycles, tend to increase due to the nature of maintenance work. There are several other reasons for the differences. Unlike production line activities, maintenance work is done under considerably harsher conditions, amid grease, grime, and dust; in the case of larger and complex equipment, such work is performed in considerably more hazardous work environments, where the maintenance crew often needs to crawl inside or underneath equipment. Furthermore, there are many complex preparations required for maintenance materials, parts, tools, and staging activities.

It should also be noted that unlike the work activities in a job shop environment, maintenance activities entail significant amounts of time spent on transit within the plant. A common feature of the maintenance crew's work is to split up into units and engage in activities designated for each unit on its own, far from supervision. This makes it difficult to assess work productivity levels for any individual case. There may well be significant differences between units that are motivated to work and those that are not.

Recently, there has been a trend to hire and train highly capable young talent as maintenance technicians. This is a definite departure from the old tradition in which the maintenance crew was trained in an apprenticeship-type environment. This trend seems to be a general and widespread phenomenon. Although personnel in the old "artisan" style may demonstrate tenacious traits with regard to the quality of their repair work, their effectiveness seems to pale with regard to scientific management of their time. The most important factor in improving maintenance productivity is to make the maintenance crew understand and appreciate the effect that repair quality has upon equipment reliability, as well as the benefits derived from preventing losses caused by stoppages. The crew must also be motivated to realize the value and significance of their work. Since it is nearly impossible to achieve direct supervision of maintenance crew members, who are typically scattered and working in many different areas of the plant, the self-motivation of each individual worker is crucial. This is a characteristic arena for management motivation to come into play.

Problems in the quality of repairs and maintenance are among the main factors that lower work productivity levels. If repair quality is poor, what was expected to last six months might cause a breakdown in machines or components within two or three months. This, in turn, causes irregularities in the intervals between breakdowns, making implementation of a maintenance plan difficult. Moreover, the maintenance department becomes increasingly reactive to unexpected breakdowns, rather than anticipating and preventing them. When small maintenance items show relatively frequent breakdowns, it is probably the result of low quality materials or inadequate repair. Thus, although direct assessment of repair quality is difficult, quality problems would eventually show up in the form of test run results or post-repair equipment problems. Evaluation of maintenance activities is difficult, and improvement in maintainability depends largely on workers' self-initiated activities and their quality of work.

4.6.2 Analysis for Maintainability Improvement

Maintainability improvements cannot be discussed without investigating the individual engineering specifics of each machine. Such investigations should be conducted based on scientific analyses and in terms of the interrelationship between the machines and their parts or components. These activities include determination of standards for diagnostic techniques, methods of specifying and maintaining the spare parts inventory, organization and assignment of maintenance units, and technical skill training. Figure 4.19 shows methods for promoting these activities.

These maintainability improvement studies should be based on reviews of the work detail records and the costs derived from the Mean Time To Repair (MTTR) analyses and the MTBF Analysis Table, following these steps:

(1) Select work activities to be investigated. Activities that formerly required numerous operational steps and high costs; those that would constitute bottlenecks in the work plan; activities that, if delayed, would have a substantial negative impact on production quality; activities that require a high level of technical skill; activities considered to be relatively low in work productivity (i.e., those that vary widely in the actual number of operational steps required for completion).

(2) Analyze the time spent on maintenance work activities. Review the steps taken to accomplish overhaul inspection and maintenance.

(3) Select the prioritized maintenance parts. Investigate the actual use records of the prioritized parts (based on results of the MTBF Analysis Tables).

(4) Investigate work methods that can reduce the time needed to do the work. These methods should include the block component replacement method, the use of prefabricated components, swift means of information exchange, loading, unloading, shipping and receiving methods, machine fixture improvements, study of synchronization of

adjustment and tuning cycles for those component groups that follow the same overhaul maintenance procedures, specifications of key maintenance quality checkpoints, survey of residual time after each job assigned, and review of the work schedules.

(5) Based on these studies, advance standardization of prioritized tasks. These include standards for operational steps of maintenance activities, estimation time standards, maintenance quality standards, spare parts quality standards, etc.

(6) Equipment improvements, study of improvements in priority maintenance parts and components, improvements in ease of maintenance work, reduction in frequency of maintenance work tasks based on synchronization of life cycles of components within the same overhaul maintenance procedures, and related study of the extension of the life span of those parts. Many topics in this category fall into engineering domains.

(7) Summarize work detail specifications by specific activities that achieved actual improvements.

(8) By combining the work detail specifications of each priority task, investigate patterns in the overall annual maintenance plan, investigate the economics of production activities, and compile the findings into an optimum annual maintenance plan.

These are procedural steps for actually implementing the improvement program. To advance the maintainability improvement program, analysis of maintenance work should be reviewed by analyzing both maintenance work hours and activities, as illustrated in what follows. The relationship between these two is shown in Figure 4.20.

$$\text{Analysis of maintenance work hours} \begin{bmatrix} \text{MTTR analysis} \\ \text{Analysis of recovery time} \end{bmatrix} \leftrightarrow \boxed{\text{Maintenance activities analysis}}$$

The first item—analysis of maintenance work hours—consists of an analysis of the MTTR and of the actual time taken to do the maintenance work. The MTTR analysis reviews the relationship between maintenance hours and frequency of occurrences, whereas the analysis of the maintenance work hours analyzes individual elements of the maintenance work and elements of activities aimed at improving them. A long-term failure analysis is a typical example of this.

Analysis of maintenance work is ultimately aimed at answering the following: From what standpoint should the maintenance depart-

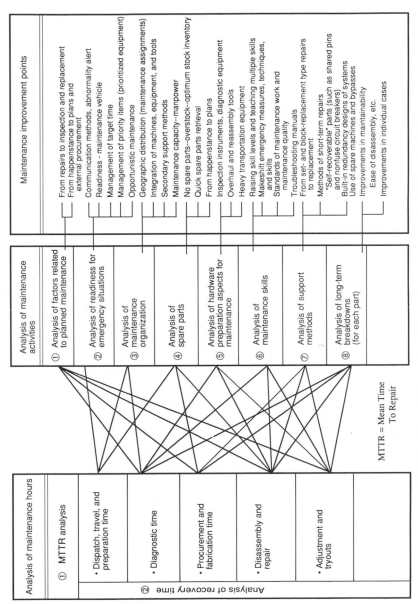

FIGURE 4.20 A Way of Looking at Improvement in Maintainability

MTTR = Mean Time To Repair

207

ment conduct the analyses in order to improve maintainability, and where are the key improvement points?

The following is an explanation of analysis methods. It should be noted that the calculation of MTTR should follow these steps:

MTTR: Mean Time To Repair
Example: 3 occurrences of failure
Total recovery time 12 hours

$$\text{MTTR} = \frac{12 \text{ hours}}{3 \text{ times}} = 4 \text{ hours per occurrence}$$

In addition to this, the 3–point estimation method can also be applied.

4.6.3 MTTR Analysis

MTTR is the acronym for mean time to repair. It is used to assess the average level and fluctuations of repair times for an entire piece of equipment as well as for individual parts or components so as to detect problems and take necessary action. An example of the MTTR analysis is shown in Figure 4.21.

The diagram of failure occurrences by repair time shows the breakdown repair time on the horizontal axis, and the number of failure occurrences on the vertical axis. As each failure is repaired and recorded on the graph, overall trends are revealed. A trend thus de-

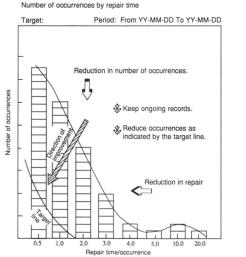

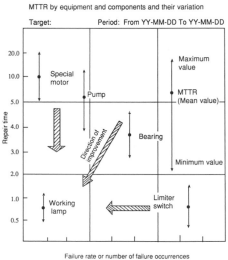

FIGURE 4.21 MTTR (Mean Time To Repair) Analysis

fined may show whether a given situation is nearing a target line that may have been determined by shop requirements, and thus aids in daily management and control. The next graph illustrates the fluctuation patterns of the MTTR by equipment or by parts, which aids in reviewing the situation in terms of zone control. It is a condensed presentation of the data for a given period of time. It shows the frequency of failure occurrences on the horizontal axis, and the repair time along the vertical axis. Each line shows an average repair time for a particular part or component, and is connected to the maximum and minimum recorded repair times for that item. This analytical diagram can aid in assessing problems in past repair work, and in setting current and future goal levels.

4.6.4 Analysis of Repair Time

For those breakdowns that took longer to repair than the target MTTR, or those breakdowns (stoppages) that lasted a long time, the time elements of maintenance work activities that actually took place should be analyzed.

The elements of maintenance work activities comprise five types: dispatch and arrival time, diagnostic time, parts issue time, disassembly and repair time, and adjustment and tryout time. For mechanical type maintenance work, the tasks deal with heavy items or difficult work characterized by long disassembly and repair time. On the other hand, for electrical maintenance work, it is the diagnostic phase that takes a long time.

Discussions of time spent in these five activities may tend to lack conceptual rigor. Nonetheless, classifying these activities and understanding their time-use characteristics can allow us to pinpoint where time was actually spent, and define the needs for reducing time spent at any particular phase. Here we can begin to clarify the need for emergency repair techniques and truly useful maintenance readiness preparation, the contents of the troubleshooting manual, and which specific maintenance tasks should be emphasized.

The following are the concepts and highlights of each of the five elements of maintenance work:

1. REDUCTION OF TIME NEEDED TO PREPARE AND ARRIVE AT THE SITE

The time frame to prepare and arrive at the trouble site begins with communication from the trouble site to the maintenance department and ends with the crew's arrival at the site. During this time, the following tasks must be accomplished:

(a) Recognition of a Failure Occurrence.

Even after a breakdown has occurred, the production crew often fails to recognize the trouble. To prevent this situation, communication methods must be established, i.e., the machine failure automatically causes the equipment to come to a halt, or an alarm bell rings.

(b) Communication With the Maintenance Department.

A considerable amount of time often elapses between a breakdown and the maintenance department's notification. This may result from long physical distances or difficulty in tracking down the right people because of some key person's absence. Thus, there is a clear need for telephone, buzzer, or predesigned indicator contact methods. Since equipment with high load levels tends to be critical to production, it is usually marked clearly on site and/or for the maintenance department. Breakdown communication about such equipment should be quick and simple. It is equally important that exactly what is wrong should be communicated as well.

(c) Arrival at the Site.

Since maintenance activities are often carried out by an emergency crew at a production site, speed of arrival is of the utmost importance. Creative measures may be needed, i.e., the maintenance crew might be stationed at or assigned to key areas where problems often occur or where breakdown avoidance is crucial. Maintenance vehicles, tools, and equipment should be kept ready. A target arrival time should be defined and maintained.

2. DIAGNOSTIC TIME

Since electrical failures are often hard to track precisely, the diagnostic phase often takes a long time. This situation also poses problems for mechanical failures. A worst-case scenario is to first dispatch the electrical maintenance crew and then the mechanical crew, one after the other, as if in a "round robin."

Experts and novices vary greatly in diagnostic accuracy. Nevertheless, using data from past breakdowns, a daily summary should be compiled noting what should be done to more quickly accomplish diagnostic cycles, define the diagnostic procedural steps, and continue the educational process. Since it is important that any individual involved may detect problem areas, diagnostic tools should be devised and other measures taken, such as installing indicator lamps or repositioning equipment. A specific period of time should be allocated for diagnostic work (e.g., 15–30 minutes), with a built-in cutoff system indicated by a buzzer, or an experienced person can be sent in support, so diagnostic work can be carried out thoroughly and without delay.

3. PARTS PROCUREMENT OR FABRICATION TIME

The time spent waiting for spare parts should be used to "batch" or fabricate them. The topic of spare parts procurement will be discussed in the section on spare parts inventory management. With regard to spare parts fabrication—of relevance only when spare parts are not in stock or available—in many cases either incorrect drawings are at hand, or there are no drawings at all, requiring replication of the part from scratch. Thus daily efforts should be made, for example, during disassembly or overhaul, to keep the drawings well organized.

4. DISASSEMBLY AND REPAIR TIME

Maintenance quality depends upon the maintenance work itself, which in turn relies on each individual maintenance worker. Depending on differences in skills and expertise, repair time may vary by as much as two or three times. A handbook of maintenance work activities should therefore be prepared to disseminate the special sense, feel, and know-how of the key points in maintenance quality.

Furthermore, in-house tools, procedures, and methods should be developed for inspection and disassembly.

Many machines totally lack design for maintainability. Machines or components that cannot be disassembled without using large and heavy equipment should be modified for easier maintenance. A study should therefore be made of set and block replacement methods for parts and components, as well as of what spare parts should be stocked, and how. Maintenance activities should be planned to reduce the number of components requiring skilled experts; most repairs should be carried out by replacement; and replaced parts should be maintained via a low-skill maintenance method.

5. ADJUSTMENT AND TRYOUT TIME

Maintenance work is often erroneously perceived as the mere repair and restoration of breakdowns. Often, tuning adjustments and tryout cannot be completed without the collaboration of the production staff. However, there is no lack of records that show the lack of proper adjustments and tryouts resulting in another breakdown soon after the repair was completed.

4.6.5 Key Points in Improving Maintainability

1. Standardization of maintenance work.
   a. Creation of a maintenance work manual. The manual should include maintenance work procedures, estimated time, target time, maintenance work quality, and key safety points.
   b. Codification of troubleshooting know-how into a manual. This manual should explain the procedural steps to investigate the problems and abnormal conditions.
   c. Improvements and clear definition of emergency repair methods.
2. Improvements in maintenance work methods.
   a. Improvements in maintainability of equipment. This includes alterations of equipment locations, installations, or structures to facilitate ease of inspection and maintenance.
   b. Creation of machine fixtures and maintenance tools. Tools for disassembly, reassembly, adjustment, tuning, inspection, etc.
   c. Availability of maintenance facilities (derricks, chain blocks, scaffolds, etc.).
   d. Change to block replacement and set replacement methods.
   e. Environmental improvements (debris, foreign matter, dirt, heat, footing, illumination, etc.).
3. Swift response system.
   a. Maintenance organization and position assignments. The questions to be decided are whether centralized or distributed maintenance should be used, and where the maintenance or stationing sites should be located.
   b. The number of maintenance crew members and task assignments, particularly job type combinations, such as mechanical and electrical maintenance.
   c. Clear definitions of communication and contact methods and dissemination of instructions, especially when the maintenance crew is not available.

   d. Facilities for maintenance such as the transportation vehicles.

   e. Changeover of the maintenance crew's work shifts (i.e., to staggered work shifts, etc.).

4. Managerial improvements.

   a. Organized maintenance of drawings.

   b. Optimized maintenance plan (including networking techniques).

   c. Concurrent repairs (taking advantage of abrupt breakdowns to carry out other tasks concurrently).

   d. Prioritization of tasks and work activities, especially with clear indications of prioritized production lines and equipment.

   e. Availability of spare parts and stocking methods (centralized vs. distributed); review of the alternatives.

5. Raising maintenance skill levels.

   a. Acquiring emergency repair techniques.

   b. Change into a multiskilled maintenance workforce with higher skill levels.

   c. Training and skill sharpening by changing group organizations into combined small work units. Changing into combined work units consisting of mechanical and electrical maintenance talent, and both novice and expert technicians.

4.6.6 Measures Against Long-Term Breakdowns

The measures taken for long-term breakdowns must always be carefully documented in a written report. Maintenance crews tend to think in terms of remedial measures that would prevent a problem from recurring. However, it could be more effective to have them think about the fastest repair methods and define a target repair time for the next possible occurrence. This approach is linked to better training of the maintenance crew. Two goals should always be emphasized to the crew with regard to long-term breakdowns: to prevent the breakdown condition and to reduce repair time. The crew should be trained to define stringent goals for future occurrences, thinking creatively in response to such questions as: What went wrong this time? Were adequate procedural steps taken for diagnostics and disassembly? Were spare parts readily available? Did the equipment structure inhibit easy maintenance? What kinds of tools should be made for future use?

The format of the mandatory report on measures taken for any long-term breakdown should include answers to the foregoing questions.

Figure 4.22 shows an example of a report on measures taken for

					Maintenance section	Production section

Report of a long-term equipment breakdown (2 hours or more)

Group name	P/S Johnson group	Line ID	MA-007		Machine no.		MI-75		Machine ID		Router				
Circumstance of Breakdown	The main shaft does not revolve due to a broken timing belt needed for rotation (the pulley shaft was rusted).		Date and time breakdown occurred			October 8, 21:30 p.m.									
			Name of failed machine or component			Timing belt									
			Time during which equipment was stopped			4.5 hours									
				Occurrence to start	Diagnostics	Staging and prep.	Procurement	Fabrication	Disassembly	Repair	Adjustment			
				0.5	0	1.5	0.25	1.0	0.5	0.5	0.25			
			Who was assigned			Maintenance		Maintenance						

Cause(s) of failure	*Why was this not detected sooner?
	The worn timing belt of the main shaft was discovered during daily inspection, and a repair was attempted on May 11, when the repair request was made to the maintenance department. Since the pulley axis could not be dislodged, the belt replacement was kept on hold. [An error in communication on this subject within the maintenance department caused a delay in its inclusion in the planned maintenance.]

Measures	Measures to prevent breakdown	Measures to reduce stoppage time
	1) Improve upon intradepartmental communication within the maintenance department. • Review order of ranking of priorities in the maintenance plan. • Since the cases as indicated in the "Cause(s) of Failure" column would cause a long-term breakdown, intradepartmental communication should indicate crucial ramifications, to avoid recurrence of this type of situation.	• Since a tool was manufactured to remove the pulley's axle, the repair time can be reduced should this situation arise again. (Keep this as in-house "know-how" and teach it to the maintenance crew.)
	2) Please conduct the belt inspection approximately once a week (production department).	If a similar breakdown occurs in the future, keep the repair time to 1 hour.

FIGURE 4.22

a long-term breakdown. An alternative report could include columns detailing the intervals within the equipment stoppage time, suggestions for preventing the breakdown's occurrence, the suggested repair time, and the financial losses incurred from the breakdown.

In examining the records of past long-term breakdowns, a clear pattern emerges of the machines and components likely to suffer long-term breakdowns. The ten worst cases could be selected and measures defined to be used in these cases. The subject of devising appropriate tools for maintenance purposes might be especially clarified by these particular situations.

4.7 SPARE PARTS MANAGEMENT

4.7.1 Goals of Spare Parts Management for Maintenance

Stock warehouses are often filled with unnecessary or noncritical supplies of materials and parts; or needed spare parts do not arrive on time, with negative effects on both maintenance and production activities. In many cases, if needed spare parts are available, a breakdown can be quickly fixed; the lack of such parts results in production stoppages and losses. Too often, measures taken against critical shortage situations may lopsidedly emphasize analysis of the causal relationships of equipment breakdowns to such a degree that investigations fail to dig deep enough to expose problems in spare parts inventory management.

Many areas in spare parts management require in-depth investigation, i.e., segmentation of management emphases, categorization of material types, determination and calculation of quantities of regularly stocked items, and methods of ordering them. In the beginning, however, it is important to remain unencumbered by excessive details. Instead, we should concentrate on such questions as why certain spare parts and materials need to be maintained and managed, in terms of their relationship to production activities. Given that the goals of PM activities are to raise the level of the equipment reliability and cost-effective maintainability, spare parts management must also deal with the purchase and manufacture of parts with a high level of reliability and the reduction of stoppage times caused by sudden breakdowns. These issues must be addressed as part of the productivity improvement study for maintenance work. In theory, so long as the number of occurrences of abrupt breakdowns can be kept to zero and all maintenance activities can be done according to a plan, parts and materials can be ordered in accordance with the needs of planned maintenance. This means that the number of regularly stocked items may be near zero.

In reality, however, the material procurement order must be placed with full awareness of the deviations of actual delivery dates that are bound to occur from time to time. For those items that are used frequently, moreover, even if their required dates and quantities can be planned, many items may be more economically purchased in bulk, keeping spare parts in stock, or an order point technique can be used whereby lowering of the stock to a certain level automatically triggers replacement orders. These procedures should all be considered for lowering order costs and more efficient management.

To seriously reduce equipment stoppage time—even for those

maintenance tasks that can be accomplished through planning—the block replacement method (an assembly type component replacement) is far more effective than the single part replacement method, and in many cases the former block assembly replacement method can assure the maintained equipment's quality. To this end, it is better to stock the parts as spare block subassembly units or components. In the case of materials and supplies for maintenance, there are consumables that are worn down or deteriorate and are then discarded, as compared to those items such as spare units or parts that are recycled. For example, large valves, pumps, speed reduction equipment, and motors can be stocked as spares. When repair work must be done, these can be replaced as complete units; the replaced units can be serviced and enter the recycling chain of restocking, replacing, and servicing. Thus there are many parts and components for which keeping a stock inventory is more economical, even though their requirements can be planned.

To summarize: The goal of spare parts management is to make spare parts stocking more economical, which in turn raises the level of equipment reliability and maintainability. With regard to cost cutting, the following points apply:

1. To cut stoppage time caused by abrupt equipment breakdowns and planned shutdowns (recovery time), regular stocking of spare parts and materials is necessary.
2. Spare parts stock management must include sufficient reviews of reliability improvements and extension of useful equipment life at the time when materials and parts are made available.
3. A method should be developed to reduce procurement and material storage costs.

Based on a thorough understanding of these points, a plan should be formulated that responds to the following questions: What and how many spare parts should be carried? With the aim of a minimum stock inventory, when and by what method should materials be procured? Where and by what method should they be stored? Who should be responsible for them? What can be done to simplify clerical and administrative handling procedures? Workers should have a heightened awareness of repair material costs to be reduced, as well as being cognizant of the fact that repair materials costs could be as high as fifty percent of total maintenance costs.

When spare parts management is analyzed, the situation diagnosed, and the problems identified, decisions relating to spare parts inventory management should depend on the characteristics of the specific plant, i.e., its location: whether it is situated in a lake region, in an industrial complex, or a rural area. If a lot of time is wasted in retrieving needed parts, if stocked parts are misplaced and duplicate purchase orders have to be issued, it is obvious that spare parts management could be improved.

A few years ago, there was debate over the relative merits of centralized as opposed to localized warehousing methods. But it is now apparent that concentrating spare parts inventory in a central warehouse to ensure inventory management boils down to an effort to avoid excessive and redundant inventories arising from sloppy management of distributed inventories. Regardless of where parts are stored, if the main consideration is to improve productivity of emergency repair work during sudden breakdowns, and if the list of stocked items and their quantities is accurate, there is no need for centralization. Since stocking spare parts is intended to reduce time needed to repair abrupt breakdowns, spare parts should be kept as close to the equipment as possible.

The level of spare parts inventory depends upon the overall level of PM activities. Obviously, more than the spare parts inventory needs good management. On the other hand, even if the level of overall maintenance is low and maintenance activities are sparse, personnel in some plants still define items and quantities to be regularly stocked and use these as a standard. In such cases, moreover, inventory bookkeeping and storage areas are well organized and kept in top condition. Such efforts are commendable, since these personnel fully understand the purpose of spare parts. In reality, however, they tend to overstock the spare parts inventory.

4.7.2 Characteristics of Spare Parts Inventory Management and Activities

Spare parts for maintenance purposes are different from parts used for production. Difficult problems exist in this area, where spare parts are characterized by a very low annual frequency of use and a slow turnover rate.

It is often difficult to determine which materials or parts should be purchased and to make plans for quantities and timing of availability. Since variations occur in the parts' life spans, moreover, even if they are periodically replaced, abrupt breakdowns cannot always be avoided. The more we try to maintain spare parts availability, the larger the regularly stocked spare parts inventory will become. The level of the spare parts inventory is therefore determined very much by the level of maintenance engineering standards and the quality of inventory management standards.

The better established a PM system is, the higher will be the probability of accurately predicting the timing of parts replacement, and the level of the overall PM promotion program will rise. If there is improvement only in spare parts inventory management, there is a limit to how much the inventory can be reduced. If we return to the basics and improve the maintenance system and techniques, a significant inventory reduction is possible. This simply means that a convoluted approach to spare parts inventory management is unnecessary if the three objectives as noted before are kept in mind. Thus, determination of spare parts inventory should take into account what items should be stocked, their quantities, and locations, and how economically and efficiently they can be maintained.

Spare parts for maintenance purposes have a high probability of obsolescence, due to equipment modifications or design changes. It is therefore important to periodically check the stock inventory. Inventory management may be complicated by a variety of factors. For instance, some parts should not be scrapped when used up, but should be serviced or reconstructed for recycling; these parts, in principle, do not need to be regularly stocked, although some should be on hand in case of an emergency. Other complicating factors relate to the form or mode in which spare parts should be stocked—as raw materials, parts, components, or in kits. This should be determined by the relationship between inventory carrying cost and losses incurred from equipment stoppages.

In light of this, it is important that the fundamental purpose of the spare parts inventory be kept in mind to arrive at the simplest form of inventory maintenance. Activities to this end can be summarized in the following three principles.

1. Management of materials at storage locations should be impeccable. This is the single most important aspect of management.
2. A simple and clear method of inventory management should be created.
3. The spare parts inventory should be reduced.

4.7.3 The Importance of Storage Location

A spare parts inventory location is judged by how quickly a given spare part can be retrieved under normal circumstances. Needed spare parts should be specified by item name. The determination as to whether a spare part is the needed one or not is made according to whether or not the item name is matched. Problems in this matching process often occur, however, resulting in the retrieval of wrong and unneeded parts. In the storage warehouse, signs and/or visual aids should be placed so that parts can be readily identified. Shelves should be easily visible, with transparent lids or covers or examples displayed, and similar parts should be clearly identified to avoid confusion.

To quickly retrieve needed parts, a comprehensive plan should be made as to which items should be kept in a centralized storage area, which in intermediate storage locations, and which should be stocked on the shop floors. For high-priority equipment, a set of spare parts should be available on the maintenance vehicles. One of the worst scenarios in inventory management is that a maintenance crew discovers that a needed spare part is kept in a padlocked storage area and the part is inaccessible. Unless each member of the maintenance crew knows on a daily basis what and how many spare parts are located exactly in which storage location, shelf, and bin, adequate inventory management is clearly lacking. If the crew cannot find stored parts simply because there is no inventory clerk, it cannot react to emergencies, a situation comparable to a gunner searching for ammunition in the heat of battle.

As noted earlier, too often the wrong size or sequence of spare parts is retrieved, or the crew may find parts in the warehouse rusting away on remote shelves, not arranged in needed sets, or stored in such a way as to make identification of usable parts impossible. The key to imaginative inventory management methods is to make the quantity of stock and its location clear and unmistakable. Measures should also be taken to prevent spare parts deterioration due to rust, scratches, dirt, and humidity. Verification should be done each time materials are replenished. The size, quality, and quantity of materials should be inspected upon receipt from vendors, and reworked or recycled items should be performance tested. Provision should also be made for retrieving and transporting the many heavy items found among spare parts.

If a spare parts storage location is observed, one glance is worth a thousand words about the level of efforts to reduce maintenance time, what parts are used for frequent breakdowns, and the attitude of the maintenance crew toward their maintenance work activities.

4.7.4 Establishment of a Spare Parts Inventory Management Method

The first aim of a spare parts inventory is the reduction of time needed to repair equipment breakdowns, especially sudden ones. If no abrupt breakdowns occurred, and all maintenance could be handled through a planned maintenance program, then spare parts would only have to be procured as needed. But the acquisition of everything through such a procurement plan would be cumbersome. On the other hand, if the maintenance crew technicians spend a sizable amount of time devising the maintenance plan or managing the spare parts inventory or materials purchasing, this is a lamentable situation. The technicians should be doing the technical and engineering jobs for which they are intended. Thus, an inventory management method should be devised that is simple, efficient, and able to be used by almost anyone. To achieve this plan, the following points should be investigated:

1. Monetary value of spare parts used, frequency of use, and cost.
2. Relationship between the inventory carrying cost and losses incurred from production stoppages arising from stock outages.
3. Should some spare parts be regularly stocked in case of abrupt breakdowns or should they be purchased according to a plan?
4. Can some spare parts be recycled?
5. What is the length of lead time needed for procurement of each spare part?

To determine this method of stock inventory for spare parts, see Figure 4.23.

During such investigations rarely used items are sometimes found, as well as excesses in the spare parts stock inventory. Questions should be raised as to whether an item should be stocked on a regular basis, by what kind of method it should be ordered, etc. The following describes the characteristics of various methods for spare parts inventory management.

1. SIMPLE FIXED-QUANTITY ORDERING METHODS

This applies to inexpensive items that are used in large quantities, such as consumables. For these items, even if the stock inventory level is somewhat excessive, it makes sense to reduce the ongoing level of inventory management. The double bin (keeping a double quantity on hand) and lot sizing methods are included in this system.

2. FIXED ORDER POINT METHODS

This is desirable if the spare parts turnover occurs every two to three months. Defining an order point will prevent depletion of the stock level below the minimum. If the inventory level of that part falls be-

Survey of quantity used

Group	(Annual usage) / (Average stock inventory)	Meaning of the values
A	More than 12	Stock depleted each month; frequent purchasing to restock.
B	4 - 12	There is always 2-3 months stock in inventory
C	2 - 4	There is always 4-6 months stock in inventory.
D	1 - 2	Although sufficient stock for 6 months or longer is always on hand, it is depleted once a year.
E	Less than 1	There is always more than 1 year's inventory in stock; no stock depletion occurs within 1 year.

Types of spare parts inventory management

No.	Order (restocking) method		Inventory status
1	Fixed-quantity ordering method — Simple methods	Double bin method • order point and order quantity are equal.	Regular stocking (regularly stocked items)
2	Fixed-quantity ordering method — Simple methods	Lot sizing • reduces administrative costs.	Regular stocking (regularly stocked items)
3	Fixed-quantity ordering method	Fixed order point method • A basic method; if inventory level drops to a certain point, an order is placed.	Regular stocking (regularly stocked items)
4	Fixed-period ordering method	Fixed-period ordering method • Inventory is checked periodically and orders are then placed.	Regular stocking (regularly stocked items)
5	Planned ordering method	Insurance stocking method • Stock inventory is kept in case of emergency.	Regular stocking (regularly stocked items)
6	Planned ordering method	Planned purchase • Orders are placed as needs arise.	Regular stocking (regularly stocked items)
7		Temporary purchase items • Orders are placed on the spur of the moment.	No stocking
8		Recycled items • Recyclable spare parts.	No stocking

FIGURE 4.23 Spare Parts Inventory Management Methods

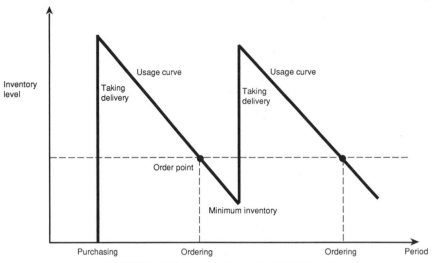

FIGURE 4.24 Fixed Order Point Method

low the order point, an order is issued automatically (see Figure 4.24). For those items with long procurement lead times, the order point should be set somewhat higher, so as to cover the long lead times between the placement of orders and their delivery.

3. FIXED-PERIOD ORDERING METHOD

In many cases, the inventory stocking method is based on criteria for keeping turnover to a period of four to six months. If this type of situation is occurring for inexpensive items, such as consumables, the stock inventory level should be reduced. Some items in this turnover category are relatively expensive in per-unit cost. For these, it is important that the monetary value of the stock be reduced. Even if stock outages of some of these spare parts may occur, if their frequency of use is not great, these outages should be of little significance. Thus, a desirable inventory management method periodically updates the physical inventory, verifies quantities on the books, discards unneeded stock, and orders only needed quantities.

4. PLANNED PURCHASE METHOD

For those items that are not used up within six months, and if the quantities used are large, the inventory method should be switched to either a fixed-period ordering method with reduced inventory quantities or lowered order points, or the fixed-quantity ordering method.

For those items that are not used in large quantities, as long as abrupt breakdowns needing such parts do not occur, replenishment periods should be determined and the parts ordered as needed based on these periods. Even for those items that are used up within six months, and as long as the planned maintenance program works properly, the inventory method for all spare parts can be changed over to the planned purchase method. This is precisely why the level of maintenance improvement can be judged on the basis of increases in the inventory levels of planned purchase items.

5. INSURANCE STOCKING METHOD

Spare parts used once a year or less are often discarded as obsolete. The insurance stocking method keeps those items in stock despite the absence of specific plans for using them. Without these items, significant losses would be incurred if and when equipment failure and production stoppage should occur, or if expensive and difficult to obtain imported manufactured products need repair. For this category, measures are needed to shorten the process of search or retrieval, and to prevent deteriorating storage conditions.

6. TEMPORARY PURCHASE ITEMS

These are items ordered after unexpected breakdowns actually occur. It is necessary that the procurement cycle be short and swift. In the spare parts warehouse, there should be a list of those parts (and their drawings), including those not currently in stock, complete with vendor information, such as contact phone numbers, cost, and purchasing lead times.

7. RECYCLED ITEMS

Recycled spare parts are regularly stocked. The key for this category is to check the actual quantities of stocked items. Figure 4.25 gives a systematic view of a spare parts inventory management system.

4.7.5 Keys to Reducing Spare Parts Inventory Cost

1. EXCESSIVE INVENTORY AND TREATMENT OF OBSOLETE ITEMS

As long as unexpected breakdowns occur, the number of spare parts tends to increase. In general, spare parts are stocked in fear of the

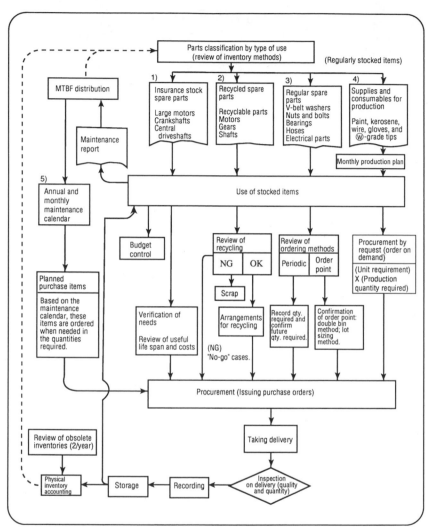

FIGURE 4.25 A Systematic View of Spare Parts Management

unexpected, leading to excessive inventory levels. Thus, the frequency of use should be investigated, and experiments made in drastic reductions. A planned maintenance program must be assiduously pursued and the life span of parts and components clearly identified. If excessive inventory for unanticipated breakdowns can be avoided, the purchase of all spare parts can be planned. Problems related to spare parts costs can be resolved through such a systematic approach, and the cost of spare parts be reduced.

2. RECYCLED SPARE PARTS

Many spare parts can be reused by recycling. For expensive items, such as those suited to fixed period ordering or planned purchases, efforts are needed to convert those parts into in-house recycled, reusable parts. Each individual part should be examined to determine how it can be recycled, where refurbishing work is needed, and where replacement is necessary. When items can be refurbished, they become recyclable spare parts, and require a different method of inventory management. To refurbish and reuse parts and components as spares, specific maintenance techniques and additional maintenance work activities are needed, but they can be quite effective.

3. PARTS STANDARDIZATION AND INTEGRATION

Machines use widely varied kinds and types of parts. However, if individual parts are examined from the standpoint of common use, they can be readily organized into groups. It is important to keep the number of parts types to a minimum. If this is successful, the level of inventory management work can be significantly reduced, as can the level of concern about stock outages. Ideally, we should be willing to cut down the number of parts types by half of what is usual today.

4. TRANSITION TO INTERNAL MAINTENANCE WORK PARTS MANUFACTURE

External contract work, or outside ordering of spare parts manufacture, is one of the most expensive categories of maintenance costs. Although there are times when external sources must be used because of the number of complex operational steps or extreme urgency, efforts should be made to convert these orders into in-house undertakings. Important equipment parts should be documented in drawings and in-house know-how accumulated about the techniques involved. For parts of special or foreign manufacture, significant future setbacks will occur without this kind of preparedness. Overhaul work on facilities is a typical example of an in-house undertaking that is very effective in terms of know-how accumulation and cost savings.

5. PROMOTION OF BUDGET MANAGEMENT THROUGH VOUCHERS INDICATING MONETARY VALUES

To instill in workers a sense of control over managing costs, vouchers are sometimes used. Regardless of expense, when abrupt breakdowns occur in equipment parts, priority is usually given to fixing them as soon as possible, forgetting the cost ramifications. The problem, however, is that this tends to be continually used as an excuse by the maintenance crew. Obviously, when a breakdown must be repaired, some expensive parts must be used. Nevertheless, this does not mean allowing loose management of spare parts inventory. To control the situation, the cost framework for each maintenance activity and production department or section can be determined, and voucher quotas can be assigned and distributed. Personnel will thus be able to manage the costs with a keener and more accurate knowledge of how much the parts and activities actually cost. Vouchering may be done via use of a unit-cost label on each part or component, or color-coded coupons may be used.

6. CARRYING OUT ACTIVITIES TO EXTEND THE PARTS' LIFE SPANS

These activities also produce significant results.

5

Self-Initiated Maintenance

5.1 BACKGROUND OF SELF-INITIATED MAINTENANCE

Table 5.1 shows a history of changes in maintenance duties over different periods. The first period was one of mainly breakdown maintenance, in which the maintenance department had a repair or taskforce squad that lacked a sense of total responsibility, or which would not initiate action until and unless it was specifically requested to do so by production personnel, even in the case of an urgent need to fix an unexpected breakdown. In such an inconvenient situation, the production department began to keep its own small repair crew. Since the department lacked the vision to develop the crew into maintenance technicians, however, there was little progress in their level of sophistication.

The second period is one of high growth, during which the number of equipment units increased dramatically, becoming sophisticated and complex. Against such a backdrop, the productivity level on the shop floor needed improvement. Two things occurred: a division of labor and work standardization. The maintenance squad, which had belonged to the production department, was moved to the maintenance department, while the production department concentrated more on the idea of "me—the person who produces." This resulted in an increased entrenchment of the division of labor. Machines were perceived as merely "assigned tools" and if they broke down, someone else would repair them. The maintenance department, however, collected technicians involved in related areas of work and became increasingly independent. This in turn raised morale and promoted a strong sense of responsibility in the maintenance crew, which began to promote preventive maintenance.

TABLE 5.1 Transition in Maintenance Work Activities

Department / Transition	Production department	Maintenance department
Period I	• The production unit has a few repair workers; they are not, however, expected to become technicians in the future.	• Repair or work (as in subcontracting) done without a sense of responsibility; or repair crew does not act unless explicitly requested to do a specific thing.
Period II	• The production unit concentrates on production, and a division of labor begins ("we, the production people, you, the maintenance people"). • The repair crew is transferred or assigned to the maintenance unit.	• Concentrates on maintenance and retains independence; keen sense of responsibility.

Period III	• The production people tend to have only limited knowledge of their production equipment.
	 • They begin to see that PM technology falls somewhere between production and processing technologies.
	 • Daily inspections are limited to simple cases only.
		• Technical and engineering skill levels need to be raised through:
		  ┌ pursuit of preventive maintenance
		  ├ diagnostic techniques
		  └ MP activities.

Period IV	• Depending on the features of each work environment, people start participating in maintenance work. There are limitations, however, due to skill levels and the number of steps in the process. • In addition to daily maintenance, workers actively participate in periodic maintenance activities as well.	• More personnel are needed for maintenance. • The production crew needs training to develop a sense of responsibility.

In the third period, the maintenance department began dealing with technical and engineering issues such as diagnostic techniques and corrective maintenance, motivated by the need to perfect preventive maintenance. However, the drive backfired to a degree, in that the production department became proportionately weak in their orientation toward equipment maintenance, a tendency that went too far. As long as the production crew kept on operating the equipment by staring at the gauges and reading the instruments, they could not be expected to improve upon their productivity level or product quality. Hence, the importance and need for incorporating PM techniques and know-how into their production and assembly techniques began to be recognized. Therefore, people began to practice some rudimentary self-initiated maintenance, such as simple cleaning, lubrication, and daily inspection.

The fourth period was a time when competition among businesses became intense, and winning or simply surviving called for the maximum effective and exhaustive use of equipment so as to cut costs, with the result that PM came to be seen as the most important thing to do.

The maintenance department began to feel the need to increase its manpower. Production personnel started to perceive that they could achieve better results by utilizing equipment effectively. They began to see that if the equipment broke down, they were the ones who would be most inconvenienced. In reality, however, they came up against constraints in the number of operational steps, as well as engineering limitations.

If we think about first-rate professionals, such as automobile racers, concert pianists, or master chefs, we realize that they use their tools or equipment—racing cars, concert pianos, or culinary implements—to help them do their jobs and reach the top of their professions. These people are deeply concerned about the condition of their tools and equipment and keep them in the finest condition. They pay attention to the state of the engine, the piano tuning, or the sharpness of their knives. They rarely rely on outside experts to check or tune their implements. Thus, to do any job really well, people must care for the tools and equipment they use and must keep both them and their skills in peak condition.

The first step in this process is to liberate the production workers, who are just beginning to become aware of PM, from skill-related constraints or limitations. The maintenance department should be responsible for training the production crew and encouraging their skills, so that they can participate confidently in maintenance activities. Instead of complaining that the number of operational steps does not give them time to do maintenance work, the production crew also should seek every opportunity and be aggressive about doing maintenance work in such circumstances. By helping with periodic and unexpected breakdown repairs, the production workers should recognize how much maintenance work they themselves can handle and always be ready to take up the next challenge.

Depending on the work environments, the production department's coverage of maintenance duties may vary. Instead of limiting its scope to simple daily service maintenance, the workers' self-initiated maintenance program should aim at improving the productivity level and product quality by teaching the mechanics of the equipment.

5.2 HIGH PRIORITY ACTIVITIES IN SELF-INITIATED MAINTENANCE

5.2.1 Equipment Problems on Shop Floors

Figure 5.1 shows data about equipment used on the shop floor where mechanical assembly work takes place. Information was gathered through collaboration between the production and maintenance groups

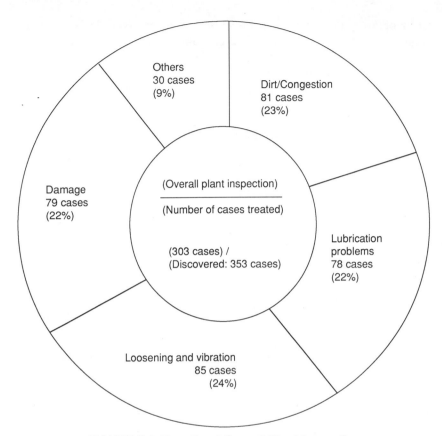

FIGURE 5.1 Results of Overall Plant Inspection

that used holiday time to come into the plant and undertake a thorough survey of the equipment.

There are three main types of problems: (1) dirt and congesting debris, (2) lack of lubricants and oil leakage, (3) loose and wobbling parts. These problems all threaten to cause immediate or inevitable brief equipment stoppages or product quality problems. All can be remedied very quickly by cleaning, lubricating or replacing lubricants, tightening screws or bolts, and otherwise adjusting or fixing the affected abnormalities. Cleaning, lubrication, and tightening are the three main activities for preventing equipment deterioration, and they can be considered the foundation of equipment maintenance.

5.2.2 What Should Be Done through Self-Initiated Maintenance

The following lists what should be undertaken as part of self-initiated maintenance. After the fifth item, the items may differ depending on specific business situations or individual workers. Levels can be raised gradually, however, as people become more familiar with the situations. In self-initiated maintenance, the three principles of "cleaning, lubrication, and tightening" should be tenaciously and thoroughly practiced. It is especially important that these basic principles be established and codified. We shall touch upon the aspects of temperature control and correct operations, which today are becoming both more important and problematic.

1. Correct operations
2. Cleaning ⎫ These three are the basic principles of prevent-
3. Lubrication ⎬ ing equipment deterioration, and the fundamen-
4. Tightening ⎭ tal conditions of equipment maintenance.
5. Daily inspection, periodic inspection, and reporting abnormalities
6. Improvement against brief stoppages and minor repairs
7. Participating and assisting in maintenance work activities
8. Spare parts inventory management, and management of machine fixtures and molds
9. Analysis of operation and maintenance records (MTBF analysis, MTBQF analysis)
10. Improvement activities

1. CLEANING

Problems Caused by Garbage and Dirt

Cleaning means removal of garbage, dirt, debris, and foreign matter, and beautification of the equipment. The following are examples of problems resulting from inadequate cleaning.

Breakdown Factors: Abrasions caused by debris on frictional clutch or rotational components, gauging or operational malfunctions caused by scum and sludge in pneumatic or hydraulic components, burned motors or contact failures in electric circuits caused by dirt adhesion, short circuits caused by interference from congested dirt on console panels, etc.

Quality Factors: Defects in precision caused by debris adhering to surfaces where something is being installed, dirt in high precision parts, inadequate electrolytic solutions caused by impurity contamination. In electronic parts manufacturing plants, clean rooms are used to fend off the onslaught of dirt and dust.

Brief Stoppage Factors: Interference in flow of workpieces caused by adhesion of dirt or debris in the ejection chutes, problems caused by dirt or metallic debris adhering to magnets or vacuum-operated components, safety problems caused by dirty phototubes. When any of these components are dirty, the inspection procedure must begin with cleaning. If the components or parts are dirty, their interiors are not visible and could provide excuses for not conducting thorough inspections.

To keep the equipment, machine fixtures, molds, and tools clean and neat aids in early detection of problems. This leads directly to the rule: "Cleaning is inspection."

Cleaning Measures

There are a considerable number of problematic situations in which the work environment may not be completely cleaned. We will review each of these situations and consider remedial measures for keeping the environment clean and neat.

Where are the places that get dirty and how dirty do they get? What kind of problems arise when they get dirty? Why do those places become dirty? Can anything be done to prevent them from getting dirty—i.e, to control the source of problems, for example, by covering those places up? What can be done to clean those places quickly and easily? What tools are needed, and where? When should cleaning be undertaken?

The specific areas to be kept clean should be pinpointed for each piece of equipment, and wiped clean during daily inspection.

> Morning inspection
> For today
> Cleaning at day's end
> For tomorrow

2. LUBRICATION

Oil-Caused Problems

There are many types of oil; here we shall deal mainly with lubrication oil and oil used in operation. At times, people drink alcoholic beverages to socialize or relieve stress. Lubricating oil serves a similar role, that of smoothing. The sites where lubricating oil, or other lubricants, are applied are the key spots in any machine's equipment. To allow burnout-type breakdowns of bearings is like failing to feed a baby milk. This is why such an occurrence is called a "shameful situation" for maintenance crews. If a PM program is in effect, the number of breakdowns caused by missed lubrication should be reduced to and kept at zero. Lack of lubrication in sliding, pneumatic, and hydraulic components or mismanagement of the hydraulic system could cause many cases of "Monday maladies."

If lubrication is inadequate and sludge accumulates in the drainage system, rust or caking could develop during holidays and weekends, manifest on Mondays in the form of operational failures. Oil starvation can also cause wear, which contributes to defects and problems with fumbled start-ups requiring adjustment work. Oil leakages may actually cause fires, and slippery surfaces may result in worker injuries.

Measures for Oils

Lubrication is needed in many places in typical machines, and even if production workers pinpoint those spots, some are apt to be overlooked. To prevent this, lubrication must be made easier, by installing a window through which to verify the lubrication condition and any changes in position, and by affixing lubrication log labels that also identify the lubrication nipples or inlets. Lubrication tubes should be checked for kinks, and oil manifolds checked to see if congestion is occurring, or if the lubricant might leak out the other end of the oil line system. Furthermore, the lubrication cycles should be extended and the amount of lubricant used reduced, through antileakage and antidegradation measures.

Even after the number of lubricants is drastically cut, and significant time is spent trying to improve lubrication application, a 100 percent success rate may still be elusive. The problem lies in reaching a thorough enough level of lubrication, let alone maintaining it. For this reason, PM begins and ends with applying lubrication.

Role of Lubricating Oils

1. Lubrication Effect ⎯⎯⎯⎯ Reduces friction
                        ⎯ Smoothes movements
                        ⎯ Prevents wear

2. Cooling Effect————————Prevents overheating and burnout
                              └— Cooling
3. Removal Effect————————Removes dirt and foreign matter
4. Prevention of Noise and——Prevents vibration and noise
   vibrations
5. Sealing Effect—————————Prevents leakage and compromised
                              pressure

It is clear from this list that application of lubrication calls for systematic techniques with significant depth.

3. TIGHTENING

What Is Looseness?

Objects are always connected to something else. The means of connection may be welding, glues, adhesives, threads, wires, or bolts. Linked objects are bound to become loosened or dislodged, resulting in a variety of problems. People can also "loosen up," and "unwind," responding by "tightening their belts," or washing their faces to wake up.

Many instances of looseness may be corrected with adhesives or welds. But in many machines, nuts and bolts are used for linkage. Thus, "tightening" is generally used to refer to prevention of looseness. Loosening may result from shocks, vibration, rocking motions, pressures, and other stresses on the machines, resulting in dislocations, axial misalignments, or shaking, until breakdowns, defects, and brief equipment stoppages occur. Many breakdowns can be traced to looseness somewhere in the machine.

Number of Looseness Situations

Table 5.2 shows the result of undertaking total bolt tightening in a plant. More than half the joinings were loose in some machines. A

TABLE 5.2 Results of Tightening

		Mechanical	Electrical	Total
Number of machines: 500 units	Number involved in tightening	35,833	233,103	268,936
	Number where loosening was detected	2,869	45,102	47,979
	Percentage of cases of loosening	8.0%	19.3%	17.8%

single production line can have hundreds or thousands of nuts and bolts, and an electrical console panel can be a veritable "nuts and bolts" monster.

Prevention of Loosening

A torque wrench should be used to tighten an important bolt. In many cases, however, "one-sided" tightening is needed, or there are "stripped" bolts caused by overtightening, or broken bolts—tightening alone is insufficient in such cases. Looseness prevention should address the following:

1. Preventing vibrations, shocks, and rocking
2. Measures to prevent loosening
3. Stopping the practice of using a single bolt for tightening
4. Using spring washers and avoiding their reuse
5. Attention to lengths, gauge thicknesses, and installation of bolts
6. Affixing matching markers where problems of loosening occur
7. Diagnosis using a test hammer

It is important that everyone work together when looking for problem areas and tightening loose joinings. In one company, all the workers joined together and went about tightening 150,000 bolts at a time. Subsequently, breakdowns were reduced as if by a miracle, and everyone learned the lesson that bolt tightening is the key to failure prevention. During a fixed amount of time when the machines are stopped, a number of "head counts" should be made. Such an approach, with total worker participation, can truly prevent failures.

4. TEMPERATURE CONTROL

Temperature exerts a variety of physical and chemical effects. Normal human body temperature is between 36 and 37 degrees Celsius (97.6 and 98.6 Fahrenheit). A temperature rise of as little as 3 to 4 degrees signifies illness. Similarly, temperature deviations in equipment cause a variety of problems. Computers, for example, are usually installed in air-conditioned rooms to protect their electronic components. Other electronic components have a maximum operating temperature limit of 40 degrees Celsius, and drastically lose function if the temperature rises. Failures in semiconductors are caused by either dirt or electric voltage, and many other electrical and electronic components lose half their life span for each 10 degrees C rise in temperature.

In a fire, the equipment's electrical wiring is often damaged, requiring major repair. In production activities, temperature-related problems include a rise or fall in the temperatures of metallic molds or galvanizing tanks having an immediate effect on productivity and product quality.

Temperature control is obviously very important, and becomes increasingly critical for equipment with electronic control mechanisms. Tools for checking temperatures include thermolabels, which allow management by observation, and heat-sensitive paints. The rule of thumb for motors is that the internal temperature should be about one-half that of the external temperature. For electrical control panels or consoles, it would be interesting to do a year's study of the relationship between internal and external temperatures. Oil tanks used in equipment operations often demand critical temperature control. The personnel who deal with them should approach their tasks with appropriate concern.

5. CORRECT OPERATION

Although operations manuals exist, most people do not read them thoroughly, often reading them only after breakdowns occur. In equipment operations there are both normal operating procedures and emergency maneuvers, differentiated by the switches and lamps on control panels and consoles. Generally, workers are only taught how to turn on the switches. But there is a need to show why such steps are needed, complete with drawings and cartoons that illustrate the equipment mechanisms and assembly principles. If the operators are not familiar enough with the equipment, they may fail to react properly when an abnormal situation arises. Thus a troubleshooting manual should be created, similar to automobile repair manuals, to help train operators. New and highly sophisticated equipment is difficult to operate. If operators do not know how to run it correctly, and make even one mistake, significant damage may result. Moreover, if operators use equipment incorrectly, irreparable damage may slowly occur. These problems are easy to describe, but difficult to solve.

Cleaning, lubricating, tightening, and temperature control are the fundamental prerequisites for equipment maintenance. The keys to successful equipment management are subtle, however, and not readily visible to all. Remedial measures against breakdowns are often undertaken without understanding of these fundamental prerequisites. Ways should therefore be devised to improve these fundamental areas through techniques of management by observation. A self-initiated maintenance program should be promoted that would guarantee proper control of the key points of cleaning, lubricating, tightening, and temperature control.

5.3 DAILY INSPECTION

Daily inspection is often practiced as part of manufacturing maintenance activities, in many cases, unfortunately, as a mere formality. We have seen many instances in which inspection goals have not been made explicit to the workers, and they merely "walk through" the checking process while following the tabulated inspection form.

Since inspection areas are usually selected to anticipate "an unlikely event," it is more than likely that even if daily inspections are carried out on the same areas, no abnormalities may be detected. This leads to careless inspections, with workers becoming oblivious to important phenomena. It is obviously annoying that regardless of neatly recorded daily inspection sheets, breakdowns occur in the very areas so checked. The purpose of conducting the daily inspections is called into question.

The purpose of inspection is not so much to detect problems in a machine as to be confident that the inspected areas—the areas "that I inspected"—can be absolutely guaranteed. It is this type of attitude that is crucial. If the inspection form has too many entry items, or if inspection standards are difficult, or if the checksheet was written well and looks fine, but it is difficult for the inspection worker to understand exactly where and how the inspection should be conducted, or if the inspection takes longer than the allotted time, it is inevitable that worker morale will diminish.

From the outset, the inspection form should neither mimic another firm's practice nor be based on some abstract concept of an ideal plant. Neither should it be too ambitious, but rather it should be in line with the level of accomplishment suited to one's work environment. After an appropriate method has been developed, the intention should be to gradually raise that level. Each worker should be motivated to develop the confidence, skills, and morale to feel a sense of accomplishment in an important undertaking.

Figure 5.2 shows the mechanics of daily inspection. Total inspection activities based on machine inspection manuals are conducted for detection of breakdowns, quality component analysis, and safety. Equipment abnormalities are divided into those that can be fixed through self-initiated maintenance and those that need intervention by other departments. The results of these undertakings are recycled in the form of one-point lessons or maintenance records.

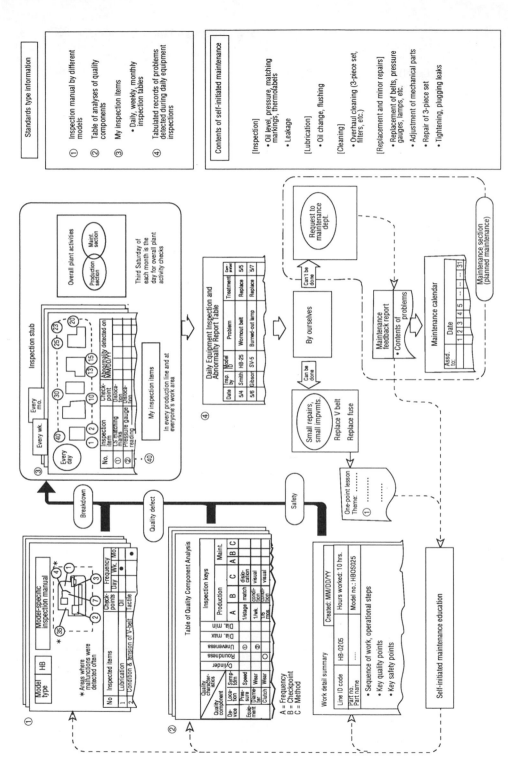

FIGURE 5.2 Procedure for Self-Initiated Maintenance

238

Selecting Inspection Items

When descriptions of items to be inspected are given, it is often unclear as to where they need lubrication. Thus, inspection items should be organized by different machine or equipment models, and a number assigned to each item in the inspection table. Workers must also be aware of why a given place must be inspected, what happens to that area when an abnormal condition arises, and what should be done in response. If a breakdown, defect, or an accident occurs, it is important that workers ask: "Why wasn't it detected sooner?"

Meaning of the Checksheet

Except for its educational role when first being used, the inspection checksheet in itself may be meaningless. If the checksheet—rather than workers' direct reports—is the only thing that can be relied on, then we would be talking about something entirely different. One should be able to say: "After the inspection is done, clean the equipment, since cleanliness is proof of the inspection's completion." This should be the watchword when instructing our subordinates, and each worker's efforts should, ideally, be respected. The intention is to change people so as to place more trust in their work.

Inspection Methods and Timing

It is often said that there is not enough time for inspection. Upon investigation, it is found that either too many items need to be inspected in a given time; the inspection methods are too difficult; or the inspection sites are hidden from view. Management by observation should therefore be employed, as in the case of thermolabels, or modifications should be made in the equipment so that inspection sites can be quickly and easily seen. The idea of management by observation for equipment boils down to a study in the productivity of inspection.

The load of activities should also be routinized, such as extending the inspection period for areas with few breakdowns, or, performing weekly rather than daily inspections. The self-initiated maintenance calendar contains more than daily inspection tables, providing for weekly or monthly maintenance activities as well.

Figure 5.3 presents an example of an inspection table for all three types: monthly, weekly, and daily. Since daily inspections must be done in a short period of time, items to be inspected are largely those that can be inspected visually or through the other four senses. Items

My inspection items

Name	John Robertson

Types of inspection items	Quantity
Lubrication--Pneumatic--Hydraulic	
Mechanical	
Electrical	
Tools and others	

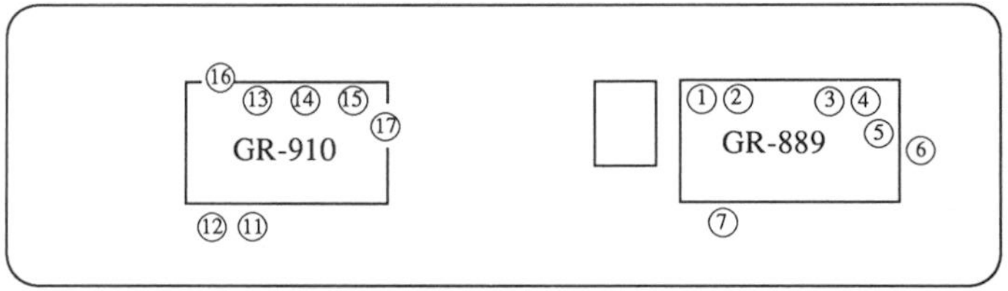

Necessary tools

WAYES-HAMMER

No.	Inspection item	Checkpoint	Date abnormality detected
Ⓠ	Matching markers on the console panel	Dislocation	
Ⓠ	Abnormal noises from spindle	Yes or No	
3	Neat and tidy organization of tools	Yes or No	
Ⓠ	Adjustment of 0 coordinate for air-micro unit	Match them	
5	Revolutions per unit of time for main shaft	Needle indication	
6	Air pressure	"	
7	Inverter instruments	"	
Ⓠ	Movement of dresser component	Movement	
Ⓠ	Matcher marker for table advancement	Dislocation	
10			

GR-910

⑯ ⑬ ⑭ ⑮ ⑰

⑫ ⑪

GR-889

① ② ③ ④ ⑤ ⑥

⑦

Ⓠ mark denotes quality component items (Q14 locations)

No.	Inspection item	Checkpoint	Date abnormality detected
11	Cooling fan abnormal noises	Yes or No	
Ⓠ	Oil Pressure Tank pressure meter	Needle reading	
13			
Ⓠ			
Ⓠ			
Ⓠ			
17			
Ⓠ			
19			
20			

FIGURE 5.3

to be inspected monthly include those that require precise and exact checking, minor servicing, and some parts replacement. Items for weekly inspection require lubrication, cleaning, and tightening.

Some personnel say that "the production people are not given much to do in the daily inspection." Figure 5.3, however, refutes this view, listing 29 items to be inspected daily. It is obvious why numbering items to be inspected in a machine makes practical sense. These should be located in easily seen places, with indicator labels attached, enabling workers to train themselves to finish the whole inspection cycle within a short period of time.

Inspection Skills

The difficult task of equipment inspection involves measuring the extent of deterioration.

The pain of a headache can only be experienced by the person suffering it. There is no way to "inspect" a headache. With regard to wear, axial misalignments, erratic movement, vibrations, and shocks—all problems begging for development of diagnostic techniques—the following questions need to be answered: How can deterioration be determined? How should it be judged? Answering these questions requires skill. If we ask workers to inspect these items without training, it is no wonder that problems in these areas will be overlooked. Once the techniques of knowing what to look for and how to look for it are taught, then each worker should carry out the diagnoses to ensure that they can do it properly.

Tools for Self-Initiated Maintenance

If inspection is considered "a necessary evil" and is, in a sense, "taken for granted," then workers would conduct inspections as well as cleaning, lubricating, and tightening bolts. For this purpose, seven tools for self-initiated maintenance should be available and carried around on the shop floor. This very specific action-oriented measure can work very well. For example, the seven tools can be attached to a belt and worn, or a toolkit may be placed anywhere on the floor.

The following are ten points to look for in developing a daily inspection program.

1. Inspection should be done at a fixed time, and items to be inspected should be regulated in terms of load and need by breaking down a process if necessary (e.g., within 5 minutes).

2. Measures should be taken so that inspection can be done visually and with ease.
3. Use tools and inspection methods creatively.
4. Clearly indicate where items to be inspected are located and their number.
5. Inspection should be carried out reliably without the aid of a checksheet.
6. Train people in skills needed for inspection.
7. Train workers so that inspection can be carried out alone.
8. Teach workers why inspection is needed, what happens if it is not done, and what happens when abnormal conditions arise.
9. Emphasize deterioration prevention more than inspection. During inspection, wipe off dirt and grime, and immediately tighten loose fastenings.
10. Always teach the value of early problem detection.

5.4 SKILL TRAINING FOR SELF-INITIATED MAINTENANCE

Lack of clarity about what exactly needs to be done is the greatest obstacle to promoting self-initiated maintenance. Too often inspections are carried out without adequate skills, or the inspection table format is used only as a formality. Workers may only know how to press a few combinations of buttons, with no knowledge even of how to remove a cover, let alone ever having touched a machine's parts, or knowing what those parts are called. Their actions are akin to driving a car without a driver's license.

Workers should study drawings of the equipment they are using. They should know: (1) What are the names of the components and parts? (2) What functions do they have or perform? (3) What kind of structures do they have? (4) What happens if any of them fail? (5) How can failures be detected? (6) What should be done about their failures? (7) What should be done to prevent them from breaking down? This is the minimum information to be learned through "one-point lessons," or whenever an equipment breakdown or abnormality occurs, by marking the relevant drawings in red ink, drawing cartoons, and the like. This is the best method of educating the workers during their daily activities.

Basic information about equipment can also be taught through group education, focusing on commonalities in different work environments. A list should be created for each shared group of equipment parts and components, thereby creating self-initiated maintenance manuals. In chapter 7, we shall discuss methods of skill training and education. For self-initiated maintenance education, a nucleus of self-initiated maintenance leaders must be established. Attempting education and training only with the maintenance crews and technicians is misguided, as well as a misuse of human resources, since it takes them away from their daily tasks. The self-initiated maintenance leaders are

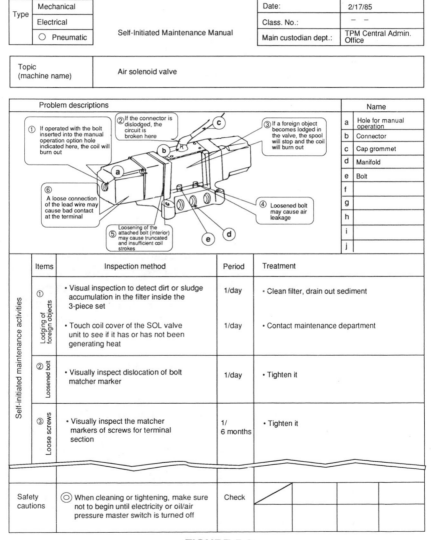

FIGURE 5.4

responsible for teaching workers skills that are applicable to their work environments and making sure that they are used.

Self-initiated maintenance education and training varies in content and level. Training and training topics should thus be grouped by different areas, and a training spectrum provided through classes of different levels, i.e., a beginners and intermediate course, or first steps, second steps, etc. Figure 5.4 gives an example of a self-initiated maintenance manual used in such training courses. It is comprised of data from all past breakdowns, which have been collected, analyzed, and organized according to component types.

At the top of the figure is a description of a past breakdown. Below it are the points and notes of caution written concerning self-initiated maintenance activities undertaken in response to the breakdown. There we see descriptions of what kind of failures the particular component is prone to have, what kind of activities should be carried out for self-initiated maintenance, what kind of tools should be used, how repairs should be made, what points should be remembered during that process—all for educational and training purposes. Either the actual table or a model should be used during training.

Workers should participate in actual maintenance tasks or periodic repair work. Since they will be dealing with the equipment that they actually use, learning can take place one step at a time as they assist in maintenance tasks. For tasks with which the students are completely unfamiliar, "hands-on" time should be available to them, together with help from the regular maintenance crews and engineering staff. They should compile a summary report of their experiences afterward, and make a presentation. This is the single most important aspect of the entire process of self-initiated maintenance skill training, and the one method that provides the best opportunity to nurture skills.

5.5 STEPS TO BE TAKEN FOR SELF-INITIATED MAINTENANCE

Figures 5.5 and 5.6 show examples of steps that should be taken for self-initiated maintenance and the mechanics of a stepwise certification system.

Goal		Step	Tasks to be accomplished
Organization of basic conditions	1.	Identification of faults and minor defects	• Implement the dictum "Cleaning is inspection." Train our eyes so that faults and minor defects can be detected while cleaning. • Take initiative by oneself and stop the practice of letting things deteriorate.
	2.	Improvements in problematic areas	• Take measures against the problem sources and eliminate problems from their roots. • Revise areas that are difficult to clean, lube and inspect, and make it easy to carry out daily maintenance activities.
	3.	Formulation of the foundation for daily maintenance	• Learn the functions and structural characteristics of the production equipment. • Establish the standards for daily maintenance routines and also learn the conditions of their application.
Analysis of weaknesses	4.	Fault diagnostics and planned maintenance	• Learn how to gather data and analyze faults. • Bring visibility to weaknesses and feedback to the designing functions. • Learn the importance and prioritization of planned maintenance.
Pursuit of how things should be	5.	Comprehensive inspection by elements	• Conduct training on inspection techniques. • Carry out a thorough inspection by elements. • Consider the measures to be taken when faults have been identified.
Self-initiated maintenance	6.	Maintenance management	• Participate in the maintenance activities that have a bearing on maintenance management and keep raising the standards.

FIGURE 5.5 Examples of Steps to Be Taken for Self-Initiated Maintenance

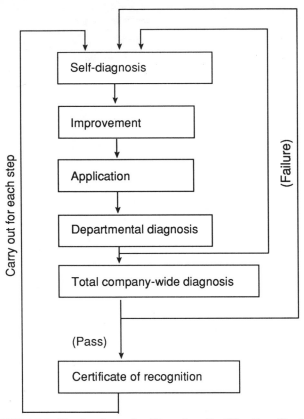

FIGURE 5.6 Mechanics of a Stepwise Certification System

6

Maintenance Prevention Design and Equipment Engineering Technology

6.1 IMPORTANCE OF EQUIPMENT ENGINEERING TECHNOLOGY

Traditionally, MP (maintenance prevention) activities have been emphasized within the framework of PM. Aimed at increasing equipment reliability and maintainability, MP has focused on the managerial aspects of PM activities. Throughout the equipment life cycle—in the design, operation, maintenance, and shutdown stages—it has sought to systematically gather and organize information. It is obvious that maintenance of existing equipment is essential in raising the level of productivity. More specifically, it is clear that thought must be given to equipment functions and structures—their reliability, maintainability, product quality reliability, safety, operability, and costs—while still in the planning, designing, and manufacturing phases, recognizing the impact of these on the subsequent effectiveness of equipment maintenance.

The relevant questions are: What kind of approach, what kind of problem definitions, and what kind of engineering technologies can production and equipment technicians use? In the following sections, we will discuss practical methods of dealing with process and equipment planning, the collection and classification of technical and engineering information, and the education and training of technicians.

6.1.1 A Life Cycle Cost (LCC) Perspective on Equipment

1. EQUIPMENT IN TERMS OF LCC

The term "equipment life cycle" is often used casually. It is specifically used to refer to problems of new, startup equipment, or to those of aging equipment undergoing overhaul and renovation—both stages having limited time spans. Equipment is rarely discussed in the context of its entire life cycle.

As shown in Figure 6.1, the life cycle cost (LCC) of equipment has two components. The first is the initial cost (IC), a combination of the cost of the planning, designing, and manufacturing phases of each piece of equipment. The second is the sustenance (SC) or running cost (RC), a combination of the costs for running, maintaining, and scrapping the equipment. LCC is the cumulative curve of the sum of the IC and the SC. It should be noted, however, that maintenance work often deals with questions of cost, particularly as to whether

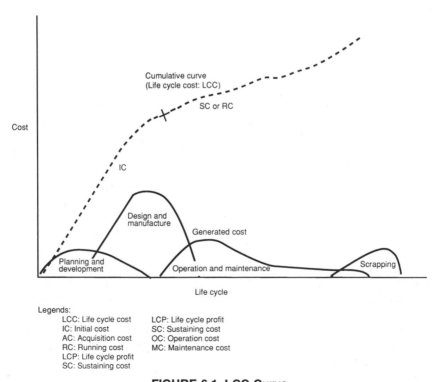

Legends:
LCC: Life cycle cost	LCP: Life cycle profit
IC: Initial cost	SC: Sustaining cost
AC: Acquisition cost	OC: Operation cost
RC: Running cost	MC: Maintenance cost
LCP: Life cycle profit	
SC: Sustaining cost	

FIGURE 6.1 LCC Curve

more should be invested in the equipment for repair work. The initial investment cost is then treated as the past or "latent" cost, attracting little attention. The common view tends to be that since the initial investment cost was high, unless the equipment continues to be fully utilized, it may generate a loss. Faced with this impressionistic view, the sense of life cycle costing is obscured.

If we look at the ratio between the initial and sustenance costs—taking domestic electric appliances as an example—sustenance costs after purchase, including electric bills, are far higher than the purchase cost. This situation is common in the case of machines or molds. In processing industries, machines have a long life cycle, between twenty and thirty years, making the share of the maintenance cost an important factor in the total LCC. In the case of molds, which require repeated investment each time product changes occur, the initial investment is of greater importance. In any event, equipment cost should first be considered in terms of LCC and then in terms of LCC's two components: initial cost (IC) and sustenance cost (SC). The cost, moreover, should be analyzed in terms of the equipment's production characteristics.

2. WHAT IS A LIFE CYCLE?

It is appropriate here to define the life cycle perspective. For human beings, the life span ends with death, in which an object, "person," ceases to function. It is a reference to a physical phenomenon. If we look at the change in a product's unit cost for a certain piece of equipment, a chart such as Figure 6.2 can be derived. At first, the proportion of costs attributable to the equipment is relatively great, causing a high product cost per unit. As the volume of production increases, however, the equipment cost ratio will decline, lowering the unit product cost. But after it reaches its lowest point, maintenance costs will increase because of losses due to production line breakdowns, or those due to defective and scrapped products, making unit costs rise. As maintenance workers have observed, equipment grows old and it becomes more effective to replace or overhaul it.

As shown in the curve's right-hand side, the physical life span of the equipment is completed when it is scrapped and replaced. When the cost curve reaches its lowest point, the equipment's economic life cycle is reached, since it is at this point that replacement of the equipment is most economical. When the equipment is replaced without relationship to either of the foregoing reasons, the replacement time is called the policy-based life cycle.

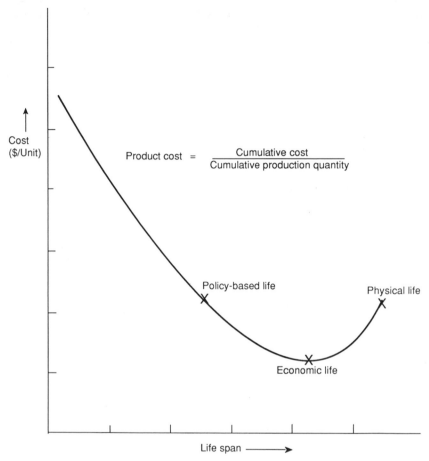

$$\text{Product cost} = \frac{\text{Cumulative cost}}{\text{Cumulative production quantity}}$$

FIGURE 6.2 Phases in a Life Cycle

In reality, however, any of these situations may be intermingled. There are many reasons for such complexities.

In the stage of equipment investment, today's usual accounting practices handle investment costs as part of production technology. With regard to costs in the operating phase, however, maintenance expenses would be charged to the maintenance department, personnel costs to the production department, raw materials to the materials department, energy to the power department, and so on. Costing is thus compartmentalized by departments, making it difficult to grasp the total life cycle cost of any given equipment. Furthermore, since no department exists to integrate the fragmented costs and evaluate an aggregate cost, adequate decision making about replacements is often neglected. Despite this, and although the maintenance department may not lack sufficient data, the secret of a practical approach to production is to lower costs during the transition period between the economic and physical life cycle periods, and to extend the total life cycle to its maximum.

6.1.2 Improving a Company's Equipment Engineering Skills

In recent years, a number of process industry companies have taken advantage of in-house development of technology and know-how about plant construction, operation, and maintenance by starting plant engineering businesses. Similarly, in machine processing and assembly industries, emphasis is being put on improving skill levels for in-house manufacturing of molds, dedicated machines, and machine fixtures, as part of overall efforts to strengthen their machine tool divisions. What accounts for these trends?

Manufacturing enterprises are engaged in development, production, and distribution of products. From the engineering standpoint, these activities can be conceptualized as a flow, shown in Figure 6.3. When technological innovations occur at the rapid rate witnessed in recent years, technological obsolescence happens faster. The cycle of obsolescence is approximately six months for product-specific technologies. This means that after only half a year, engineering competitive advantages will be significantly reduced. Although the obsolescence cycle for equipment-related technologies is about eighteen months, neither a production technology nor a maintenance technology can be easily copied by another firm. Thus, if we build equipment in which this type of technology is interdependent, the end product can better compete with other firms, instead of being subject only to the technological obsolescence cycle.

No matter how good a product may be, its advantages may be eclipsed by substitutions of processing methods, conditions of fabrication, and/or conditions of operation. Those changes, in turn, will be interpreted as a revised set of equipment conditions and will subsequently be embodied in another generation of equipment. This simply means that products are always produced through use of equipment, and the differences in equipment engineering are directly reflected in differences of productivity and competitive power—including quality and costs—among the products of various enterprises.

Despite this, equipment engineers too often merely send for product literature and, without adding anything of their own, "shop around"

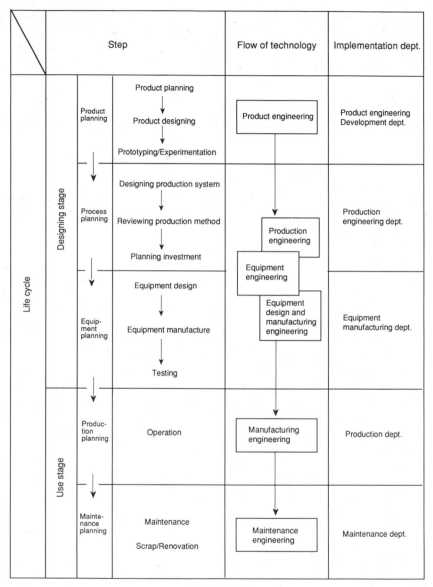

FIGURE 6.3

for products that seem most suited to current known needs, and begin constructing production lines on those assumptions. This could well be called "catalog engineering." As long as engineers continue to rely only on purchased technologies, no matter how advanced they may be, money cannot buy truly competitive equipment. Although there may be some differences in timing, competitors will also obtain these technologies. It is crucial to remember that the source of a plant's competitive power springs from differences in equipment engineering prowess. This is painfully apparent to all businesses, and is precisely the reason why all strive for internally owned technologies and in-house generated know-how.

In-house creation of equipment and production technology, and thorough investigation of the equipment's reliability, maintainability, and operability, can make or break productive maintenance. Thus, it is highly desirable to improve the skills of the production and equipment engineering technicians. Accumulation of equipment engineering know-how and construction of equipment with technologically unique characteristics is an essential part of the corporate assets of any manufacturing enterprise.

6.1.3 Laying the Foundation of Equipment Technologies

To date PM activities have concentrated on measures against breakdowns, or corrective maintenance, in the maintenance department, and in improving quality and operations in the production department. Most activities in these areas are responses to shoddiness and equipment problems. They are tantamount to aftercare of poor equipment planning, designing, and manufacturing stages. Personnel involved in such "aftercare" have tried to have their feedback included in the next generation of equipment. The maintenance and production departments have reported PM information, while asking that production and equipment engineering technicians not make the same errors twice. In hindsight, their efforts were largely a one-sided mode of communication.

Against a background of ever-increasing sophistication, complexity, and overall integration, those products or equipment aimed at a mass or undifferentiated market—i.e., automobiles, electric home appliances, machine tools, and boiler units—rely on product quality as the main selling point. Manufacturers make major efforts to perfect product quality management, and refrain from distributing products until they have confidence in their ability to hold their own in the marketplace.

Products and parts come in different shapes and sizes, with different material characteristics. Unless unique in-house production techniques and know-how have been developed, some products may not be able to meet quality or cost requirements for a given situation. The equipment may be geared to small-volume or individualized production modes, or be pilot models with many unknown factors as potential problems. Despite this, equipment engineers still try to develop efficient equipment to meet customer requirements, taking new ideas as well as operational aspects into consideration.

From the standpoint of the operations and maintenance departments, even if there were many successes and only one problem arose, the complaints received might sound as if everything was a failure. But forcing the engineers to dwell only on failures would lower their morale. When the question arises as to who is to be blamed for a problem, the group providing the equipment and the group operating and maintaining it find themselves in opposite camps. It is no wonder that they do not agree. In some companies, the engineers are the ones to take the full brunt of criticism. The engineers' position depends entirely on the firm's attitude. Rather than having engineers who are perpetually afraid of failures and walk only on the safe side, we need to nurture those engineers who would set difficult goals and respond to challenges. By doing so, at the very least we can anticipate plans that will take us to the outer limit of what can be done, even if they mean only higher product reliability. In such an environment, those engineers can be groomed to be free of conventional concepts, able to use new and original ideas, and with the spiritual fiber to continue making productive innovations.

Thus it is important that the various issues are treated not merely as concerns of equipment engineers, but as issues related to laying the foundation to cultivate in-house-developed equipment know-how, and mobilizing everyone in equipment engineering—including production, maintenance, and industrial engineering. Maintenance and production engineers should unite in cultivating, accumulating, and increasing equipment engineering know-how. These goals should be reflected in PM activities with the ultimate aim of creating equipment that is user-oriented, provides user satisfaction, and is uniquely suited to the particular work environment.

6.1.4 Problems During Initial Equipment Management Period

With a few exceptions, personnel in many industries are not adept at creating their own in-house equipment. In many cases, firms simply specify the operating conditions and leave things up to the engineering technologies of the equipment manufacturers. Or they rely mainly on selecting ready-made equipment and then buy engineering customization contracts. As a rule, these firms are engaged in supervising the equipment's routing, which includes layout and plumbing design, and tasks of construction and installation.

When the time comes for live testing, however, problems arise and normal operation does not take place readily. Especially in machine processing and assembly industries, the operation and maintenance engineering staff is confronted by major difficulties. In many cases, modification after modification is needed before the plant or its production lines can begin normal operations. Even after this stage has been reached, equipment operability may be so bad that it is difficult to conduct the routine inspections, adjustments, lubrication, cleaning, and minor repairs crucial for extending the equipment's life cycle and preventing breakdowns. At times, the maintenance crew may have to crawl underneath a machine just to make a small adjustment, and minor work may cause work clothes to be completely soiled by grease or oil.

When we analyze work done for operational and maintenance improvement, it is often found that if workers had only given more thought to the initial survey analysis and design stages, the problems would have not occurred (see Figure 6.4).

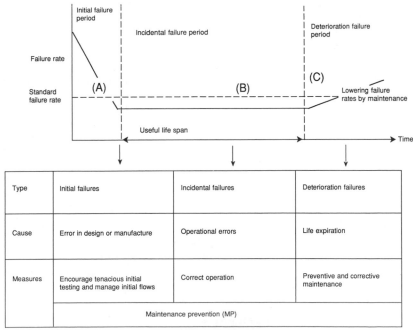

FIGURE 6.4 Characteristics of Life Span and Measures Against Breakdowns

6.1.5 Engineering Data and Know-how for Design of Reliability and Maintainability

Engineers are generally afraid of making mistakes. Before designing anything, they study past technologies and add their own, sometimes revolutionary, ideas. In process industries, equipment is installed for twenty to thirty years; hence, the design stage is critical. In industries where products are constantly changing, unless the firm can adapt its equipment, it is doomed to lose its competitive edge, which also makes the design stage crucial. From an engineering standpoint, given different engineering requirements and specialized areas, engineering expertise is not easily acquired. What then should engineering division managers do to efficiently groom a sufficient number of young engineers to meet these needs?

An engineer's growth is significantly affected by his or her creativity and self-initiated efforts—the "willingness to do," expressed in an attitude that says: "let's see it, let's think it out, let's try it." For design and maintenance engineers (the authors among them), occasions may arise when we doubt our own abilities, but we will never fail to find subjects to study and topics to challenge us.

Theories developed on the desktop or drawing board are not the only contributions to technological innovation. Many lessons are learned empirically. These lessons should not merely remain individual learning experiences, but should be organized by engineers within the firm and used to train personnel in a corporate perspective.

Engineers make mistakes daily and often express their frustration by never being quite satisfied with what they have done, always lamenting as if to say "it should have been done this way" or "it would have been made better had it been done that way." At the same time, they cannot afford to repeat the same mistakes. Rather, engineers should learn from the mistakes of others and incorporate the lessons learned in their own work. One approach involves collecting engineering data that can be used for designing reliability and maintainability, and including them in engineering design standards.

Design engineers are typically swamped by new construction and expansion, with little time for reflection. They may therefore leave considerations of design improvement to the maintenance engineers. We believe that it is important that both design and maintenance engineers together create the data for engineering standards for equipment reliability and maintainability. These should be so interesting that the design engineers cannot help but review them before designing or selecting equipment.

Passive feedback of maintenance information to the design engineers is ineffective. Rather, the maintenance engineers should feel that they are backing up the design engineers. And the design engineers should have a continuing sense of responsibility for the equipment, sparing no efforts to follow up. Moreover, they should think about designing solutions to problems that have arisen in the past. If the communication gap between the design and maintenance engineering groups can be bridged, if a corporate engineering know-how handbook is compiled through a joint effort, if the engineering data are continuously accumulated in an organized way, it would not only greatly contribute to the education and training of younger engineers, but would also contribute to the building of corporate engineering assets not easily surpassed by the competition.

6.1.6 Product and Equipment Costs

It is time to sort out what is meant by costs, including the difference between product and equipment costs. As can be seen in Figure 6.5, the product cost consists of labor, equipment, material, and other costs. Reducing each individual cost within the product cost reduces the total cost. However, this has a profound effect on the way equipment is built. For example, if we compare a manual and automated manufacturing line, both of which make an identical product, it is seen that the manual manufacturing line incurs low equipment costs while incurring high labor costs. This can be described as a production line with low fixed costs. On the other hand, the automated manufacturing line would incur low labor costs and high equipment costs. Since both production lines have the same product cost, we cannot immediately tell which production method is better. The question of how much the equipment cost should be relative to the product cost, however, or how the equipment should be designed, should be considered in terms of market trends and other factors. That is, the product cost depends very much on equipment engineering. Thus equipment engineers believe that it is the equipment engineering that determines the product costs.

A second point concerns equipment costs. As indicated in Figure 6.5, the initial cost is often an overly impressive figure. In businesses, the accounting department deals with the return on investment as part of investment budget management. The endemic problem, however, is the accounting department's tendency to consider the return on investment only in the context of the initial investment funds. Despite

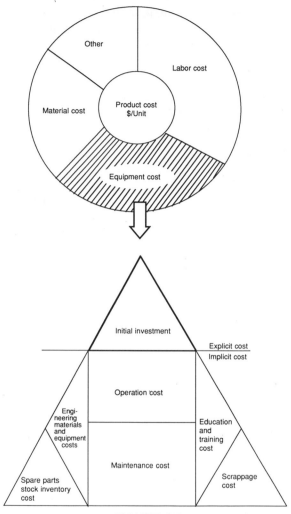

FIGURE 6.5

many cases in which maintenance costs far surpass the initial cost, maintenance costs tend to lose their visibility because they tend to increase by small increments over an extended period of time.

Why do these equipment maintenance costs occur? Let us consider the example of costs for equipment upkeep. If a machine must be repaired after a sudden breakdown, the question of reliability is raised. Furthermore, if defective products were produced, and molds and machine fixtures needed repair, additional material costs due to inferior materials and lost processing costs would be added to the usual costs of upkeep. These highlight the point that unless reliability and maintainability are fully considered, inadequacies in these areas will profoundly affect both product and equipment costs.

The engineers' attitude toward their tasks is also relevant here. The return on investment is an index derived from a statistical calculation, which the accounting department understands. To improve return on investment, however, is an engineering matter, determined by how well and skillfully engineering tasks are accomplished. Engineers tend to think that if only there was more money available, a better method could be chosen. This typical thinking reflects their tendency to shy away from cost consciousness and the thought of cost constraints. While making strenuous efforts to design functional aspects, they tend to fall short in designing for the life cycle cost of equipment, although the latter is equally important. Although LCC aims to minimize the total cost, engineers should extend their costing horizon beyond the life cycle of a given machine, and define their lifetime goal as designing the most profitable equipment possible.

6.2 THEMES FOR EQUIPMENT ENGINEERING

How can progress be made in accumulating equipment engineering know-how and training younger engineers? We propose four activity and task zones, as shown in Figure 6.6. In the upper left-hand quadrangle is a zone (Zone I) for creating equipment, including planning, designing, and manufacturing. The lower left-hand quadrangle (Zone II), shows activities performed throughout the equipment life cycle, beginning with initial material flow management, through operation, maintenance, and scrapping.

The lower right-hand quadrangle (Zone III) represents activities aimed at gathering engineering information from equipment building, operation, and maintenance, and its inclusion as a corporate know-how asset. The upper right-hand zone (Zone IV), shows ways of raising engineering skill levels. It is here that people deal with methods of recognizing and evaluating specific goals, and the specific engineering tasks by which to implement them. These are supportive procedures to improve equipment engineering skills, and are thus directed to equipment-building tasks.

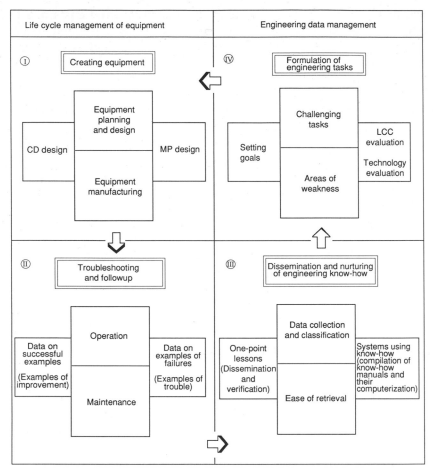

FIGURE 6.6 Themes of Equipment Engineers' Activities—The Four Zones

6.3 SYSTEM FOR IMPLEMENTING AN EQUIPMENT ENGINEERING PROJECT

Figure 6.7 gives an overview of how equipment-building activities can be carried out. The approach is to see the procedures as part of a system for implementing an equipment engineering project. In the middle of the diagram is the equipment-creating activity block, initiated by the flow from the product diagram, which moves through process planning → equipment design → manufacturing → trial runs. This block is interfaced with another flow block—an equipment usage phase—that moves through initial flow control → production → main-

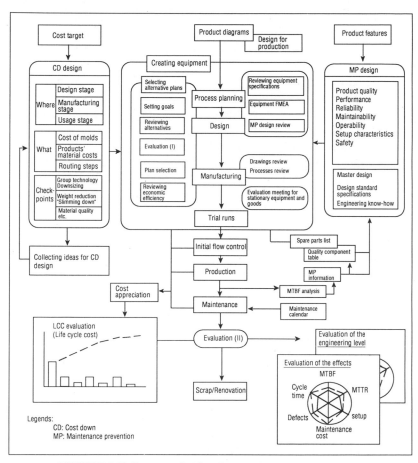

**FIGURE 6.7 System for Implementing an Equipment
Engineering Project**

tenance → scrapping. These comprise the entire life cycle of a facility
from its beginning to its end as scrap.

Equipment is created in steps that include the role of engineers in
determining goals and reviewing and assessing alternatives. In this
process, the engineering pros and cons and economic factors are considered. These techniques are used throughout each step of engineering method evaluation, process planning, equipment designing, and
equipment manufacturing. In reviewing the practices of different business firms, we noted that alternative plans are sometimes presented
only as a formality. Moreover, some alternative plans are so contro-
versial that their plausibility could only be judged if they were put

into practice. What is important here is the accumulation of experience and know-how, training, and engineering expertise that allow good plans to be generated. Good plans cannot be made without creation of such an infrastructure.

It is important to distinguish designing for CD (cost down) and for MP (maintenance prevention). There are many situations in which the two contradict each other, giving designers only limited freedom and providing them with excuses for not meeting the given design requirements. Designing for CD begins with one of its goals: to meet a given target product cost. It is important that the designers are given responsibility to design the cost itself. It would be useful to give them a compilation of examples and ideas in this regard. Beyond cost considerations, designing for MP aims at designing equipment that meets user requirements. The alternative plans for CD and MP designs should be reviewed and adjusted, and a choice made as to which will ultimately be adopted. Before the final equipment is installed in a production plant, it is usually tested during a trial period. A pilot plant may be created, and while tests are conducted, engineering know-how is acquired. When this phase ends, additional consideration should be given to aspects of supportability and maintainability. Such efforts may include providing a specification list indicating key areas or weak components, or a list of spare parts to be stocked.

When production is about to begin, an initial material flow management is undertaken. Remedial measures are then examined for the new equipment's start-up "bugs." Recently, however, a "live" production start-up has been demanded at this stage, and the target level is thus set significantly higher. In reality, there still remains a gap between this high level of expectation and what the new equipment can provide. Hence, losses at initial start-up are increasingly becoming an issue of major concern.

When equipment begins production, the MTBF Analysis Tables are created to record reliability and maintainability, and daily operation logs, problem reports, and examples of remedies are used as well. The data from these are fed back to the design department as the MP information. Modifications may also be made in the spare parts list. The equipment is kept up according to the maintenance calendar and, to grasp the cost ramifications, the cost-related activities are logged in the cost accounting ledger. The latter serves to evaluate the project from a technological standpoint, through the life cycle costing (LCC) evaluation, the assessment of the equipment's effectiveness, and the assessment of the project's engineering level. The findings are used to set future directions and improvements. The foregoing can be summarized in nine steps.

Step 1. Definition of goals

Step 2. Creation of alternative plans using CD design and evaluation techniques

Step 3. Creation of alternative plans using MP design and evaluation techniques

Step 4. Economic efficiency review and assessment

Step 5. Compilation of findings into an equipment engineering assessment table

Step 6. Assessment of initial tryouts and material flow management

Step 7. Maintenance recordkeeping method and MP information collection

Step 8. Implementation of equipment maintenance plans based on maintenance calendar

Step 9. Life cycle costing and engineering assessment

Steps 1, 2, 3, 5, 6, and 9 will be discussed in what follows.

The design stage determines most equipment costs, functions, and effectiveness (70 to 90 percent). Goal definition, the most important step, means that design engineers make a specific commitment about a project area.

Some goals are harder than others to attain, and some are known to be attainable from the start. Even if engineers set only one goal, it should be clearly identified, with the intention of being reached. To do this, the most important aspect of the project must be defined. Furthermore, market trends for the intended product must be understood, so that equipment is appropriately designed.

Within the constraints of costing plans and marketing considerations, the engineers should determine product costs. With regard to cost details, the reader is referred to section 6.1.6.

Promotion of the life cycle cost (LCC) is intended to minimize the total LCC, which consists of the initial equipment investment plus the equipment operation and maintenance costs. Today, it is true that for all industrial enterprises, products are diverse and demands unstable, making it difficult to build production equipment. Such being the case, it is preferable to keep the initial equipment investment as low as possible. There is a general trend, however, whereby equipment is built with many extra functions. This suggests to us that in the context of LCC, the most urgent goal is to reduce the initial investment.

Table 6.2 categorizes the items of assessment for evaluating equipment functions and effectiveness. These should be made operational by quantifying each item.

As a yardstick for defining goals, we have chosen here to look at the results obtained for similar pieces of equipment. This is done because the production line is the stage in which the LCC is maintained and used. Since most such assessments are meaningful in the production lines, the true picture can be disclosed in its entirety by observing similar equipment. This method also allows the engineers to have first-hand experience in studying the real requirements of real production environments.

Some engineers are afraid that if they define goals, they will be held responsible for them. Others feel that goals should be only loosely defined. To promote a relaxed atmosphere among the design engineers, other engineers from the manufacturing, production, and maintenance departments should join them in determining goals. Everyone should clearly understand the target costs and the kinds of equipment desired by each department, and each engineer must understand exactly what his or her department's tasks are in the total project, as well as individual assignments. Unless these steps are followed, the final equipment will be lacking in both tangible and intangible ways, unpopular, and hence not likely to be used with care.

6.4 KEY POINTS IN DESIGNING FOR COST DOWN

Just as designers must be concerned with function, they must also be aware of costs. What happens, however, if designers continue to lower safety indexes and shave costs? On the other hand, if functional constraints are too great, implementing ingenious design ideas is too difficult. To avoid this, design engineers should forget about various constraints and emphasize design ideas at first. Notwithstanding the presence of functional constraints, there is plenty of room for reducing costs without touching the functional aspects.

Designing for CD (cost down) can be approached in three ways.

The first approach is based on the idea that product cost is to be anchored to the equipment engineering technology. This approach examines new engineering technologies, processing plans, manufacturing methods, and production designs, as well as production equipment automation. It attempts to deal directly with reduction of raw material and labor costs (see Figure 6.8).

The second approach derives from the idea of creating equipment based on cost cycle considerations. The underlying thinking is as follows: Once the investment is made, equipment costs remain as fixed

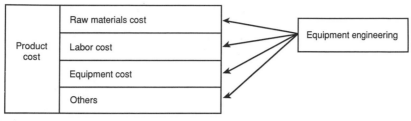

FIGURE 6.8

costs. A mistake in the initial phase, therefore, is irrevocable. On the other hand, material costs are variable, and do not increase each time an additional unit is produced. They can be cut as soon as they are found to be unnecessary for running the production line.

With regard to the initial investment cost (IC) and sustenance costs (SC): Even if the LCC remains constant—depending on which side the cost may keep increasing—three situations can be considered, as shown in Figure 6.9. The demands of the product pose a major question for the equipment life cycle, i.e., for how long should the equipment be run at what production rate?

Type I (Figure 6.9), is a general purpose machine, as exemplified in its equipment and processing or milling machines. These can fabricate many different products as long as the appropriate machine fixtures, bits, and molds are in place. It is a machine designed to be flexible and to accommodate a variety of accessories, with a high level of reliability, and a long life span. This machine's weaknesses are that setup time to change accessories tends to be long and adjustment time tends to be significant as well.

Type II is a dedicated machine, apparent in its molds, machine fixtures, and a kind of dedicated production line, made specifically for certain product(s). Its design objectives are very clear, making possible automation and elimination of unnecessary functions. Its weaknesses include a lack of general applicability, which means that if the products become obsolete, so does the equipment, even if it is still usable. For this type, profits must be made within a short period of time.

Type III uses consumable items, such as lathe bits, tips, and bearings. These are replaced at certain intervals to produce certain types of products. Spare parts belong to this category.

The above typology can be applied not only to the overall equipment and production lines, but also to molds and machine fixtures. If it is applied to the way the latter are made, significant cost reductions can be achieved. The main point is that if we reduce areas within

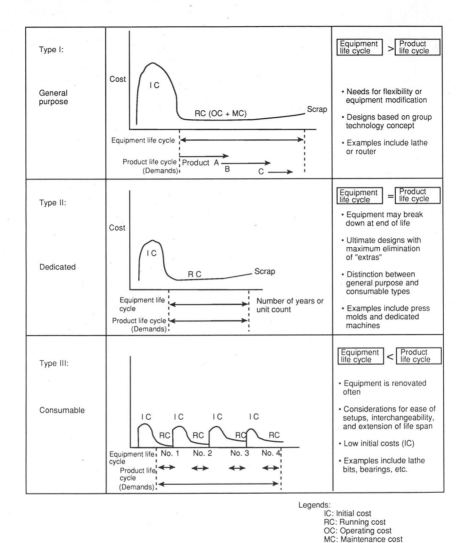

Legends:
IC: Initial cost
RC: Running cost
OC: Operating cost
MC: Maintenance cost

FIGURE 6.9 Three Types of Equipment Planning and Life Cycles

equipment that are dedicated to special and limited applications, we can raise the level of reliability and achieve general purpose equipment. In this regard, parts that are to be changed should be as small, local, inexpensive, and as simply changed as possible.

The third approach is the design of a cost tree. The cost tree can locate ways to reduce costs by first breaking down the equipment, step by step, to its components, units, or parts. At each step the equipment cost is analyzed per unit so that the target and actual costs can be compared. Design engineers may have been content to simply design with a safe margin of error. Once exposed to the cost tree approach, however, which seeks the ultimate in cost design, then, and for the first time, the question of reliability truly emerges. This is in stark contrast to an easygoing decision-making process based on intuition or a "seat-of-the pants" method. In implementing the cost tree approach, it is often found that much of today's equipment has undergone little cost scrutiny or examination, and, moreover, is not user-oriented.

Obviously, we should seek to define the real needs of the shop floor, and create equipment to uniquely meet those needs. Questions should be asked, such as: What can be gained if this equipment is used? Will it be inexpensive? Will it better handle material queuing? Will it reduce the number of operational processing steps?

To reduce waste through designing for CD and to make user-oriented equipment, the key points have been sorted and shown in Table 6.1, using an example of this approach for a mold. Based on the viewpoint shown in moving in the direction of the arrow from the left-hand to the right-hand side of the table, we believe that a new way of thinking about equipment building will emerge. It is possible that cases may exist in which the direction of the arrow is reversed.

6.5 KEY POINTS IN DESIGNING FOR MAINTENANCE PREVENTION

Some people may be very good at a narrowly specialized area, but once outside that area, they may be helpless. Engineers tend to concentrate in their areas of specialization, and to overlook what they are not interested in. An analogy can be found in equipment, in that if a completed machine is carefully examined, the areas of strength of its designer are immediately apparent. Designing for MP means including the idea of preventive maintenance from the initial design stage, to avoid making the same mistake twice. Table 6.2 shows the key points in designing for MP. Building the MP concept into the very design means assuring product quality, reliability, maintainability, operability, setup ease, and a comfortable environment. This should be the design engineer's most important task. Too often, once the initial stage shows that the product meets its requirements, other details are likely to be neglected, and little information about the sustenance and operation phases will be available. This is tantamount to ignoring issues of reliability and maintainability.

MP information should be categorized and compiled for existing

TABLE 6.1 Design for Cost Down (CD) and Checkpoints (Examples)

Step	No.	Checkpoints	Examples	Initial running costs	
				I	R
Process planning	1	Process → Development of new processes; Process substitution	Development of new techniques (process methods, equipment) Press → roll, bender → roll press	○	○
	2	Number of processes → No process, reduced, combined, complex	Based on design changes for value engineering (VE) reasons, parts are designed in common groups; use of different functions for no-step processes; combination of processes or process steps (build into a roll press); use of complex cam gears to allow total multiassembly processes with one mold and one stroke.	○	○
	3	Dedicated → General purpose, common usage	Combined use based on one-touch adjustment of a punch	○	
	4	New design and installation → Transfer New design and installation → Alteration of modified parts. Use of spare machines	Use of experimental models Recycling	○	
	5	High speed → necessary speed (synchronized with required cycle time)	Function-priority models with no extras or frills (Resiliency)	○	○
Machine planning	6	Automatic → Semiautomatic		○	
	7	Power source → air pressure/oil pressure → mechanical/other power	From cylinders to motors Mechanical devices using other power sources	○	○
	8	Large size (heavy) → compact (light)	"Cassette-ization" (contents replaceable) Miniaturization based on functional priority principles	○	
	9	Complex → simple		○	
	10	Long life span → Necessary life span	Lowering of material grades Use of "fire-sale" punches	○	
Design specification reviews and approvals	11	High-grade materials → lower-grade materials	Materials matched with production quantity and cycle time (Rockwell hardness 83 → 81) Use of plastics	○	
	12	Many parts and components → few parts and components		○	
	13	Large capacity → smaller capacity		○	
	14	Time-consuming setup requirements → short setups	Die sets must be integrated with the machine or integrated with modularized machine fixtures; use of tap-load/unload types		○
	15	Time-consuming supply replenishment types → short supply	Use of ball-retainer punches, cassettes		○
	16	Time-consuming adjustment types → short adjustment time types	Use of the press-bender type adjustment techniques	○	
	17	Use of unique and specialized parts → Common standard parts	Use of off-the-shelf parts	○	
	18	Large number of safety margins → Designs with calculated limits		○	
	19	Precision for all possible process usage → precision within actual needs		○	
	20	Fabrication methods by rules → multiple choices	Arcs, wire cutting, misjudgment → use of sanders	○	
	21	Individualized designs → automated designs; standardized drawings	Computerization; use of the American Manufacturing Standards (AMS)	○	
	22	Vendor-dependent corrections → in-house corrections	In-house work matching body components, safety guard implements	○	

TABLE 6.2 Key Points in Designing Maintenance Prevention (MP)

Product quality	Design to ensure product quality and manufacturing precision
Functionality	Design with consideration of equipment functions and speed, etc.
Reliability	Design to prevent equipment breakdowns
Maintainability	Design to allow quick recovery from breakdowns
Operability and ease of setups	Design for easy operation and setup
Safety	Design to consider built-in safety features
Environmental pleasantness	Design to incorporate the 5S's for a good work environment

problems and measures to avoid their recurrence. Using an MP information table, these data are fed back to engineers during the design phase. It is especially important that this information be given a standardized presentation in charts and diagrams. A compilation of problem reports may also be useful.

Conventional wisdom suggests that if we design for P (production), and try to build a good piece of equipment, then the equipment cost will rise sharply. Furthermore, the design review and materials procurement will take significantly longer. This is true to a great extent, since if an error is made somewhere, then it would have a side effect horizontally in that residual functions may be tacked on to unrelated pieces of equipment. It is also true that if there is a lack of prior investigation, corrections may have to be made after the equipment is completed, requiring greater efforts than if they had been designed into the original plan after careful prior reviews. In this systematic designing concept, this is one reason why experts advocate the practicing CD design without being encumbered early on by MP designing constraints. After waste has been eliminated through the use of the CD designing technique, however, it is still necessary that each item for MP design be accounted for.

Specifically, the MP design review document should be created for real past problems and future anticipated ones, enabling remedial measures to be available for predictable problems. Figure 6.10 is an example of this approach. It includes assessment of factors contribut-

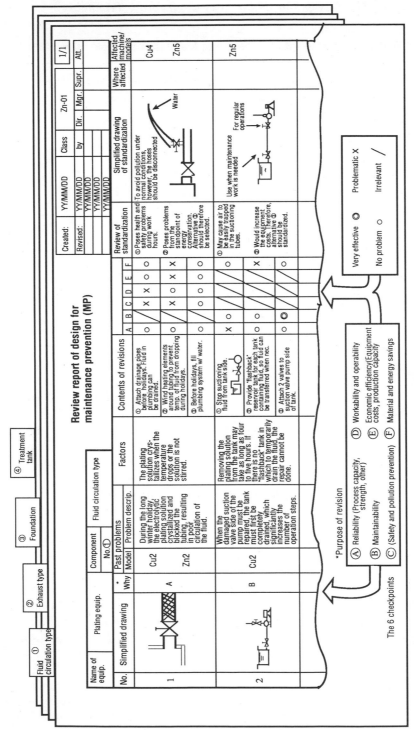

FIGURE 6.10

TABLE 6.3 Reliability Design Checkpoints

1. Establish clear standards for outside purchase of parts and materials, such as inspections upon receipt or certification tests.
2. Minimize number of parts used and maximize simplification of their structures.
3. Simplify assembly equipment and increase assembly reliability.
4. Minimize number of areas that need servicing and enable inspection and/or testing of each workpiece as assembled.
5. Provide for ease of maintenance.
6. Use good parts. If mass-produced, dependable parts are used, the equipment's reliability level should be high.
7. Set aside sufficient reserves of crucial parts and components.
8. If necessary, use redundant methods (e.g. parallel; standby).
9. Devise safety measures against stress.
10. Analyze design reliability at both initial and final stages.
11. Eliminate human factors as much as possible.
12. Conduct reliability tests.
13. Define procedures to test design prototypes and to eliminate defects in trial models.
14. When modifying designs, follow strict monitoring procedures to prevent breakdown recurrences.
15. Structures needed for maintenance:

    1) Measurement tools for internal inspections; manholes and peepholes.
    2) Passageways, staircases, and functional utility equipment: ladders, scaffolding, hoisting equipment. Plan space needed for maintenance.
    3) Provision for removing fluids used in maintenance and disposal of accumulated deposits.
    4) Plans for internal rinsing and steaming, etc.
    5) Plan for ease of inspection and replacement of internally used parts and components.

ing to problems, revised plans, and their evaluation. Evaluation is done in terms of the MP categories, such as reliability and maintainability, and based on the cost and degree of urgency. Those items that should be standardized should be the key points for design.

MP design review is done similarly, through each step of process planning, design, and manufacture. In addition to the MP evaluation document, other review documents are created, e.g., to meet the equipment's FMEA (failure mode and effects analysis) and other requirements, with the same basic approach carried throughout.

As an alternative to the evaluation document, know-how accumulated from past experience can be compiled into a master design or an engineering standards document. A checksheet may also be useful. Table 6.3 shows an example of key checkpoints.

Even if a good checksheet or PM system may be in place, some companies may still have multiple problems. Unless equipment engineers review their equipment, their activities remain a mere formality.

In the MP design stage, the equipment's spare parts must be reviewed. This process can take place only after the design engineers specify the weak components and assess their life cycle. Off-the-cuff guesswork or trial calculations are not good enough—past data must be used as reference. Thus, the maintenance crew must review how the data are collected to give advice on maintenance and engineering matters and truly help the design engineers.

6.6 PROJECT SUMMARY AND EVALUATION

Notwithstanding variations in the extent, every manufacturing firm is involved in the practice of equipment engineering. However, data are often found to be in total disarray in many different and fragmented files. Statistical data are often categorized and maintained by different functions, thereby confusing even design engineers. For these reasons, a table of related data should be devised for each project undertaken, a single glance at which would give an overall perspective of the project. An example of this is given in Figure 6.11. A look at the table shows project goals, targeted values, and the revision plans and their effects. This table facilitates action and thus provides important data for management.

Project evaluation seeks to verify if project goals are being met, and if any action needs to be taken.

Two approaches are needed for evaluation. The first is a cost evaluation using the LCC concept, and the second is an evaluation of the effect, or impact, of the project. Although business must ultimately be measured in terms of money, unless its output is improved, cost reductions alone cannot raise a project's value. Hence, evaluation of a project's effects can be helped by providing columns in the diagram for assessing engineering work, so that the rise in the level of engineering work can be shown to have contributed to cost reduction.

6.7 INITIAL MATERIAL FLOW AND MP INFORMATION MANAGEMENT

1. INITIAL PRODUCTION FLOW MANAGEMENT

Recent trends show a tendency toward shorter product life spans and a demand for compressed production start-up times. Unless money is made during the start-up period, the chance for a profit may be lost; hence, the importance of initial production flow management. This

Table evaluating life cycle cost (LCC) of a mold — July 4, 1983

Production engineering dept.	Equipment manufacturing dept.	Production dept.	Maintenance dept.

Product name:	Similar product:	Customer	T automobile corp.
S1010Y	S1000X		S1000X

Goals of LCC

Since the production quantity is small, there will be only a marginal return on investment if the conventional method is used. The R and L models should therefore be combined into one.

Conditions

Customer		T automobile corp.
Product life span		4 years
Production plan	Monthly average	15,000
	Peak month	20,000
Expected startup date:		Oct./1983

MP design

Type	Ideas
Quality	Improvements in precision of clearances of punches and dies
Reliability	Surface treatment of punches, TIC
Safety	Preventing dropping of stopper
Reliability	Ultrafine finish of the surface "push" section R

Simplified drawing

No. of alternative plans:	30 cases
No. of plans adopted:	20 cases

Cost down design

Type	Ideas
I	Combined use of R and L types
I	Change in thickness of die set from 150mm to 50mm
R	Material change for punches and dies from SKH9 → SKD11
R	Improvement in yield

Simplified drawing (150 → 50)

No. of alternative plans:	20 cases
No. of plans adopted:	15 cases

Overall evaluation

Evaluation items	Similar accomplishment	Target	Ratio	Actual (Trial)	Actual (Initial)
Material cost	$53	$49			
Manufacturing cost	$62	$50			
Investment amount	$2,000	$1,300			
Production cost	$132	$121			

Cost / Initial period / Production runs

	Evaluation item	Similar products' accomplishments	Target	Rate	Accomplishments Trials	Initial	Rate
Cost	Material cost	$500,000	$200,000				
	Manufacturing cost	1,350	900				
	Trials and adjustment costs	100	50				
	Equipment alteration cost	0	0				
	Spare parts cost	50	30				
Initial period	Cycle time (minutes)	15SPM	20SPM				
Production runs	Setup time (minute/minute)	30	10				
	Defects rate (%)	0.08	0.01				
	Yield (%)	68	80				
	MTBF (shots)	4,800	10,000				
	MTTR (minute)	35	10				
	Type of life cycle (shots)						

LCC curve

Evaluation chart — MTBE, MTTR, Maintenance cost, Defects rate

Contact with maintenance and production departments

Differences between goals and accomplishments

FIGURE 6.11

effort is worthwhile, even if all the firm's engineering resources are necessary.

Initial production flow management falls into Zone II of equipment engineering. It is the stage that precedes mass production. The purpose of management at this stage is to run the equipment with the designated number of staff, materials, and methods, detect initial problems early on, and eliminate them. In preparation for this stage, production operators are trained in skills and operational steps, and the maintenance crew is trained in emergency measures and spare parts procurement. The equipment is specified as being in the initial production flow management phase. This specification remains until production reaches a certain predesignated level, at which time it can be lifted.

2. MP INFORMATION MANAGEMENT

Problems that may arise while the equipment is being run-in or maintained, cases of improvement modifications, or examples of measures taken are all engineers' "treasures," providing valuable information. This data should be channeled as the MP information, and should be used for taking measures to avoid repetition of problems. A clear, tabulated MP information sheet should be available to the design engineers, enabling them to act readily as well as being stored as basic engineering data. Its format and method of retrieval are therefore important. Figure 6.12 gives one example of a tabulated format, as well as showing key points for managing an MP system.

3. KEY POINTS FOR MANAGING AN MP SYSTEM

1. Conduct a thorough investigation at the time of equipment completion and at the time of initial production flow management with regard to aspects of the equipment's quantitative and qualitative processing capability (see Figure 6.13).

   Investigation results should be compiled in a report of the equipment's Process Capability (CP), and fed back to the planning, design, and manufacturing departments, and the subcontractor who made the equipment. When the equipment reaches either the time for overhaul, or a certain standard production quantity level, then an investigation should be conducted to accurately assess the process capability (which refers to the qualitative capacity of the process). If the characteristics of the manufactured product are expressed in terms of a calculated value, then an index of pro-

Item no.		Product type	Type class	Goals of revisions				Problems	Treatment
Item name	Various hinge nails	☐ Cover	☐ Transfer	Impvt.	☑ Quality stabilization	☐ Safety		☐ Warpage	☐ Adjustment
Dept.	Production	☐ Diaphragm	☑ Sequential advancement		☐ Operation rate	☐ Pollution		☐ Crack	☑ Correction
Model no.	PR 0041	☐ Disk	☐ Acquired material engineering	Reliability	☐ Durability	☐ Environment		☐ Scratch	☐ Replacement
Line ID		☐ Door lock	☐ Choking		☐ Extended life span	☐ Energy Saving		☐ Waviness or wrinkles	☐ Rework
1 ☐	Reference information	☑ Door hinge	☐ Outer shape formation engineering		☐ 5S characteristics	☐ VA•VE		☐ Size aberration	☐ Shape
		☐ Booster	☐ Molding	Impvt.	☐ Inspectability	☐ Workability/ Operability		☑ Damage	☐ Material alteration
2 ☐	Action needed: / Desired completion date: YY/MM/DD	☐ Jack	☐ Drilling	Maintainability	☐ Lubrication serviceability	☐ Standardization		☐ Chipping	☐ Surface finish
		☐ Body parts	☐ Cutting		☐ Ease of disassembly and replacement	☑ Parts Cost		☐ Bending	☐ Label indication
3 ☐	Implementation: / Desired completion date: YY/MM/DD	☐ Moldings	☐ Other		☐ Interchangeability	☐ Other		☐ Other	☐ Other

Description of problem

During continuous processing of various types of sequentially conveyed hinge nail models, debris would too often become lodged in the core of the punched-out cavity area causing cracks on various parts of the mold.

Remedy

Removed chuting ducts attached to the mold, drilled a hole in the press bed, and removed debris from under press.

Illustration (photograph)

Illustration (photograph)

Calculating value of the remedy   Part cost: 150,000 x 12 months   180,000/year

Comments by the production engineering department:

Future specifications will be made to minimize stray air stream that contributes to clogging.

Comments by the design department:

Since the air stream can cause unexpected trouble in that area, we will adopt the method described above as much as possible.

Maintenance section			Production engineering section			Equipment design section			File
Manager	Supervisor	Attendant	Manager	Supervisor	Attendant	Manager	Supervisor	Attendant	

FIGURE 6.12

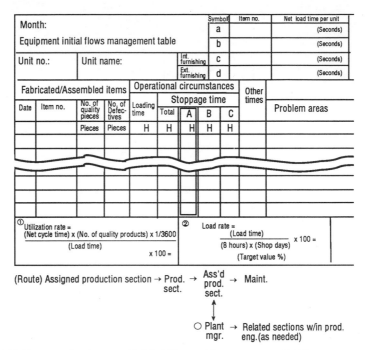

FIGURE 6.13 Equipment Initial Flows Management Table

cess capability (CP) is used; if they are expressed in terms of a ratio, then it refers to the defect rate (compare with P).

Investigation of the process capability requires an understanding of the four elements of which it is comprised: equipment, materials, personnel, and processing method. In order for this to happen, we need to first clearly identify the items that need to be investigated in light of the four elements, and also investigate the method of sampling the products and the method of measurement. It is especially important to understand how equipment functioning or deterioration has affected product quality.

In order for the product, process design, and equipment planning departments—as well as the production and maintenance departments—to take full advantage of the process capability, there is a need for information based on a complete engineering analysis of the total production flow. This includes analyzing components, identifying measures to be taken, and making proposals for improvements. It is important to ensure that data derived from these investigations are used in the report.

2. The design and maintenance departments as well as the end-user

○ Stoppage time A: Stoppage time caused by failure of this equipment and quality defects. B: Stoppage time caused by setups, fixture changeovers, adjustments, and daily inspections. C: Stoppages caused by other production activities. ○ Other times: Stoppage time due to machine idling caused by component shortage, or equipment shutdowns caused by power-related problems, and stoppage time due to causes other than production activities authorized by company management.	Plant manager	Production engineering section			Production section		
		Section mgr.	Supervisor	Attendant	Section mgr.	Supervisor	Attendant
	Initial flow started on: YY/MM/DD				Department		

Nature of the problems of and corrective measures for the cause of A				Result of daily inspection table (A)			Number of operations for total output	
Remedial actions	Work done by:	Follow-up results		Lubrication	Cleaning	Inspec-Adjust-ment	Total no. of Operations	No. of Operations for total output

③ Scrap rate =

$$\frac{\text{(No. of defectives)}}{\text{(No. of quality products) x (No. of defectives)}} \text{ x 100} = \text{(Target value \%)}$$

④ Frequency rate =

$$\frac{\text{(Failure occurrences) (No. of occurrences of A)}}{\text{(Load time)}} \text{ x 100} =$$

⑤ Resiliency rate =

$$\frac{\text{(Cumulative down time) } (\Sigma \text{ A})}{\text{(Load time)}} \text{ x 100} = \text{(Goal \%)}$$

[Notes]

1. Utilization rate

2. To investigate and calculate load rate, see "Plant Operations Survey Manual"

should jointly conduct a periodic investigation of the prioritized equipment use; review the equipment's design and manufacturing aspects; and collect the valuable information for MP activities.

3. To ensure that information obtained from analysis data or collected by direct investigation is used, or that it is fed back from other areas into the next cycle of process and equipment designing, the information should be entered in individual equipment history files, either under production lines or equipment. Engineering information that is common to many pieces of equipment should be accumulated and become part of the machine design standards document (subdivided into reliability, maintainability, safety, etc.). In this way, it will be accessible to all the company's equipment engineers, and can also be used as training material.

4. The life cycle of each piece of equipment should be studied in the context of its dependency on the environmental conditions of prioritized common parts or components, and engineering data should be collected for improvement in component reliability.

Two main breakdown patterns can be identified. One is an abrupt or sudden breakdown of parts such as delimiter and solenoid switches, cylinders, valves, and seals. Study of these should focus on their life spans. The second breakdown pattern concerns functionally deteriorating parts, e.g. guide bushings, wave pattern chutes, cams, and cam follows. Study of these should be done periodically and focus on understanding how they wear out.

6.8 BUILDING KNOW-HOW AND TECHNOLOGICAL TRAINING

Both problems and solutions are strikingly similar among different companies. Does this mean that engineers are too busy to review past troubles? Or does it mean that novice engineers tried to deal with problems and made errors? The pace of technological progress continues to increase, accompanied by product diversification and integration, confronting engineers with ever tougher problems. But why do engineers play "cat and mouse" with engineering innovations to such an extent that it would appear that no progress has been made at all? And what should we be doing in such a situation?

Considerable originality, ingenuity, and innovation are seen in many manufacturing enterprises. But the problem is that these activities are not built into corporate engineering assets, nor are sufficient efforts made to do so. Regardless of informational content or ideas, companies should think in terms of how information can be accumulated as the firm's corporate engineering know-how and systematically used as its asset. These relationships are shown in Figure 6.14, and can be explained as follows: The first step is concerned with raw data and actual cases. At this stage, the most that can expected to be accomplished is compilation of a casebook. It should be noted that these data, which are mostly result-oriented, are not necessarily organized in ways engineers find useful. Moreover, there may be many very similar cases represented, and hence continuing data collection may be of little use from an engineering standpoint. Ironically, the more this kind of data is collected, the less useful will its collection be— until it becomes totally useless.

The second step is to organize the material as engineering information, compiled as a collection of engineering know-how. At this stage, the data are categorized for engineers' use. Each case is to be examined in terms of "know-how" and "know-why"—i.e., in terms of principles and concepts—and organized accordingly. This collection is clearly intended to help engineers find new ideas and suggestions when used.

The third step is to refine engineering know-how into the firm's own corporate engineering standards. Once these become the stan-

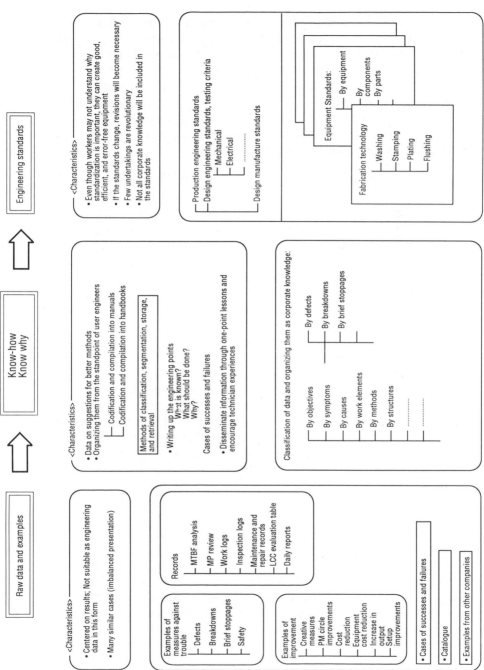

FIGURE 6.14 Engineering Know-how and Standards

Raw data and examples

Know-how
Know why

Engineering standards

<Characteristics>
- Even though workers may not understand why standardization is important, they can create good, efficient, and error-free equipment
- If the standards change, revisions will become necessary
- Few undertakings are revolutionary
- Not all corporate knowledge will be included in the standards

Production engineering standards
Design engineering standards, testing criteria
├─ Mechanical
├─ Electrical
└─ ....
Design manufacture standards

Fabrication technology
├─ Washing
├─ Stamping
├─ Plating
└─ Flushing

Equipment Standards:
├─ By equipment
├─ By components
└─ By parts

<Characteristics>
- Data on suggestions for better methods
- Organizing them from the standpoint of user engineers
  ├─ Codification and compilation into manuals
  └─ Codification and compilation into handbooks

Methods of classification, segmentation, storage, and retrieval

- Writing up the engineering points
  What is known?
  What should be done?
  Why?
- Cases of successes and failures
- Disseminate information through one-point lessons and encourage technician experiences

Classification of data and organizing them as corporate knowledge:
├─ By objectives   By defects
├─ By symptoms   By breakdowns
├─ By causes   By brief stoppages
├─ By work elements
├─ By methods
├─ By structures
├─ ....
└─ ....

<Characteristics>
- Centered on results; Not suitable as engineering data in this form
- Many similar cases (imbalanced presentation)

Records
├─ MTBF analysis
├─ MP review
├─ Work logs
├─ Inspection logs
├─ Maintenance and repair records
├─ LCC evaluation table
└─ Daily reports

Examples of measures against trouble
├─ Defects
├─ Breakdowns
├─ Brief stoppages
└─ Safety

Examples of improvement
├─ Creative measures
├─ PM circle improvements
├─ Cost reduction
├─ Equipment cost reduction
├─ Increase in output
└─ Setup improvements

- Cases of successes and failures
- Catalogue
- Examples from other companies

279

dardized rules to be followed, novice engineers can become more efficient and avoid mistakes. Standardization procedures are especially indispensable in Computer-Aided Design (CAD), Computer-Aided Manufacturing (CAM), and Computer-Aided Engineering (CAE) activities. Not all the engineering know-how, however, would become part of the engineering standards. Sometimes, too, the question becomes obscure as to why the given standards are important, and innovative approaches may be discouraged. Moreover, standards are bound to become obsolete and need revision.

Building Know-how

Engineering knowledge is an organized body of accumulated experiences that have been refined and screened through the mesh of commonly accepted principles and axioms. The way in which these experiences are categorized and segmented, however, is an ongoing issue. Since such knowledge is of little value if it cannot be retrieved when needed, storage and retrieval measures are crucial. It is also necessary that knowledge or information be readily disseminated, which can be accomplished in two ways.

The first method provides a database for storing information. Since each piece of data is classified in such a way that it must be retrievable from multiple viewpoints, the complex problem of "multipoint" retrieval cannot be solved unless computers are used. If the data are categorized and stored on computer, retrieval becomes relatively easy. Thus, the database must be established to allow for different kinds of retrieval and to be user friendly. Otherwise, the system would be soon abandoned. Although this type of system looks good on paper, it has many problems; much progress remains to be made in computerized office automation equipment.

Another approach organizes the data as a collection of engineering know-how from the engineer's point of view. This means classifying the data according to chosen categories, collecting information for each, and setting up a filing system. Since it contains only the kind of information needed, the file should not be very large. Its greatest advantage is that engineers can take the initiative in reviewing the data, and, while organizing them, can create a collection of know-how. In this case, engineers are likely to handle it with care and respect. Time should be found once a year to organize the engineering data in this way.

6.9 ORGANIZATION OF ENGINEERING TASKS

From the engineer's perspective, most of what happens on the production shop floor is often seen as predetermined. In other words, unless engineers have visions and dreams about the future, no progress or innovation will take place on the shop floor. On the other hand, the question must also be raised as to whether engineers are given the kind of environment that is conducive to having visions of the future or the time to think about them. If engineers are running around at work from morning until night preoccupied with troubleshooting, they have no time to develop aspirations for the future. But the engineers are also responsible for the way in which they work. Unless engineers attain a deeper understanding of reality and are clearheaded about what their challenges and tasks should be, they will fall victim to the "frogs around the pond" syndrome. Hence we had engineers get together and discuss what they view as desirable, as shown in Figure 6.15.

It is now important to clearly identify engineering tasks, according to the following steps:

Step 1: Sort out current problems and arrange in order.

Step 2: Organize issues by engineering tasks and determine their levels of importance.

Step 3: Create the task map.

Step 4: List tasks.

Step 5: Solve the problems by task.

Step 6: Determine goals of the engineering standards and their evaluation.

Step 7: Write the engineering summary and review.

In step 1, current problems and tasks are organized. Here, comparisons should be made between present and future and between the company and its competitors. Survey papers should be written about the current situation of the production technologies, engineering equipment, and maintenance, listing the problems and tasks by a focus on specific product groups or process characteristics. Figure 6.16 is an example of this approach.

In step 2, these problems and tasks are organized from engineering viewpoints. This involves two operations: the first is to review topics on engineering weaknesses, which should summarize the tasks for which there are no good solutions at present. The second is to

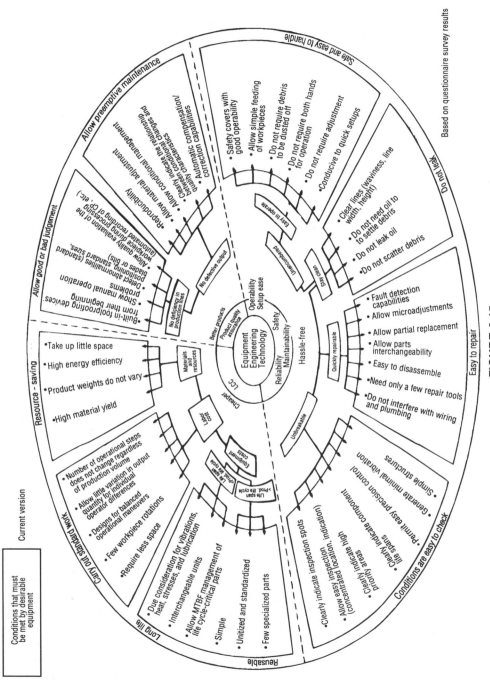

FIGURE 6.15

Based on questionnaire survey results

Conditions that must be met by desirable equipment | Current version

282

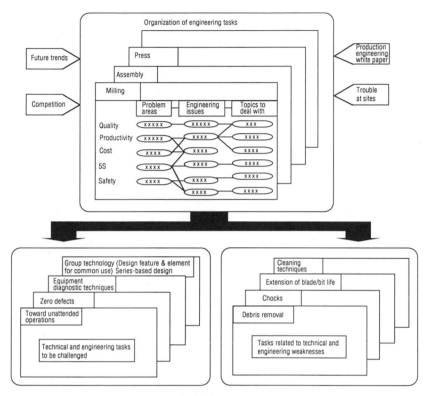

FIGURE 6.16

compile those tasks that comprise engineering challenges, i.e., those not yet dealt with in-house, but where involvement should take place. In the first case, for example, the life span of a lathe bit might be extended from the current 500-piece revolutions to a 5,000-piece revolutions. An example of a second case could be to make improvements in unattended operation from a rate of 10 hours of operation per defect one year from now to 24 hours of operation per defect two years from now.

In step 3, problems are organized in the form of task maps or tables that are quickly and easily understood.

In steps 4 and 5, problem themes are selected and listed, and target dates are set for solution.

In step 6, engineering standards are evaluated for these tasks. This important step is intended to steer the engineers in certain directions.

It should be done each time an equipment engineering project is undertaken, so that engineering standards can be evaluated.

In step 7, the pieces of information that led to problem solution are collected as corporate know-how. It should be noted that this know-how and the engineering tasks are in a reciprocal relationship.

7

Equipment-Knowledgeable Personnel—Skills Training

7.1 ENGINEERING AND SKILLS NEEDED FOR EQUIPMENT

When complex and highly sophisticated equipment, production lines, or molds are introduced, are they likely to operate efficiently from the very beginning? Most staff, maintenance crew, and production technicians would say, "No." Most likely, production efficiency at start-up would be less than 50 percent. If it rose above 70 percent, the equipment manufacturer's representatives and the staff might "freak out" and leave the site wondering what was happening.

The start-up process is generally difficult and slow, since production equipment is comprised of both developing and fixed elements. Depending on the companies, there may be a wide range of products manufactured and equipment used, and know-how has not yet been accumulated.

What kind of skills are needed to make effective, maximum use of the equipment? Figure 7.1 illustrates what we think those requirements are. Equipment design is equipment engineering. Thus, training engineers who have knowledge of equipment means training engineers in equipment design. In equipment manufacturing departments (for fabrication machines, for example), we see a recent trend where companies prefer to make the molds and dedicated machines in-house. By doing so, they expect to accumulate engineering and technical know-how as corporate assets. The process of developing manufacturing skills is considered an especially crucial undertaking. Considering the varied requirements of manufacturing shop floors, it is important that equipment design and manufacturing engineers develop equipment with

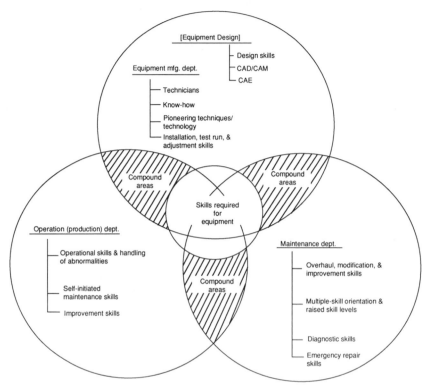

For compound areas, rotations or overseas training/education may serve the purpose.

FIGURE 7.1 Skills Required for Equipment

unique characteristics that are enthusiastically accepted by production personnel on the shop floor. It is at the design and manufacturing stages that the equipment's basic functions are determined.

The maintenance department plays a central role in the effective use of equipment. We know of situations where recovery from breakdowns is delayed simply because the malfunctioning equipment has too many electronic components, and other situations where overhaul or renovation resulted in improved equipment function and precision.

The production department's role is obviously important as well. Operational errors in highly sophisticated equipment, or the inability to deal with abnormal conditions, lower the equipment utilization rate. Progress must be made toward improvements in the setup, innovations in foolproofing devices, increase in processing speeds, or automation for unattended operation in order to operate equipment with a high utilization rate while maintaining product quality and tight cost control.

The weight given to each of these three areas—equipment design, maintenance, and production—differs from industry to industry. In process industries, for example, once equipment is installed, organized maintenance department skills become very important; whereas in machine processing and assembly industries, self-initiated and innovative maintenance skills play a decisive role. The engineering and technical skills required also vary, not only according to differing equipment's characteristics and products, but by departments as well. Intradepartmental cooperation is necessary to train personnel in equipment know-how and to meet the needs within each department.

7.2 TRAINING PROGRAMS

7.2.1 Transition to Multiskilled Human Resources and a Skills Inventory List

Production lines in the past were organized in terms of specific jobs or types of processes performed, and machines were often scattered on the shop floor. The recent trend is to setting up production lines with increasingly diverse, complex, and interdependent machines and equipment. A variety of skills is needed to operate a modern production line, and a weakness in any one of them may cause a significant loss to the business.

Using Figure 7.2, let us examine the required skills inventory list. These can be broken down into equipment operation, maintenance, inspection, and other types of skills. We can identify the type of skills needed at each job or workstation, i.e., at each production line and in other areas of the work environment. Seen in this way, it is apparent that for each skill there are matching jobs, such as production, maintenance, and inspection work.

The next step is to take an inventory of individual personnel with the skills needed at each workstation—production, maintenance crew, etc.—to identify where certain skills are missing, and where and how they are needed. Planning for the appropriate combination of matching skills constitutes multiskill orientation/training.

The teaching of technical skills and specific engineering techniques can still be divided into traditional specialized areas, such as mechanical and electrical engineering, with a trend toward subspecialties. On the other hand, as in the case of mechanical and electronic engineering merged as "mechatronics," a wider integration of some engineering fields is taking place. On production floors, moreover, a number of problems arise only because some workers do not have multiple skills.

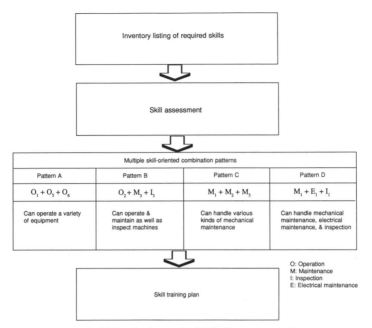

FIGURE 7.2 Steps in Skill Assessment

How then can levels of individual specialization be raised while simultaneously imparting a broader spectrum of knowledge? Can appropriate sets of multiskill patterns be taught? In parallel, what kind of innovations should be introduced in the way equipment is made and the methods by which it is maintained?

Let us examine the way combinations and patterns of multiskill sets can be defined. In Figure 7.2, the first case (O1 + O5 + O6) denotes a situation in which a worker can economically operate various kinds of equipment, and can thus be used in a number of production activities. Moreover, it allows workers to cooperate by either offering or receiving assistance without difficulty. The next combination (O2 + M3 + I3) denotes a situation in which a single worker can do both operation and maintenance. Although the work may entail specialized skills, it takes place in a process industry environment. The third case (M1 + M2 + M3) denotes a situation in which a maintenance crew member can perform various kinds of tasks within one job class, i.e., various mechanical maintenance tasks. The last example (M1 + E1 + P1) denotes a situation in which the worker can also perform mechanical, electrical, and plumbing maintenance.

7.2.2 Guaranteeing Maintenance Quality

Equipment-related skills include manufacturing, maintenance, and operating skills; maintenance skills will be discussed here. Maintenance skills depend on three factors: speed, accuracy, and safety. Maintenance work affects production. Maintenance activities include maintenance skills, diagnostic techniques, tools for disassembly and reassembly, and compilation of manuals—but maintenance skills are of primary importance. Maintenance quality—the ability to complete maintenance activities without error—is determined by maintenance skills (see Figure 7.3).

Repair quality is neither easily explained nor understood and is quite different from product quality. The latter is quantitatively defined, i.e., product features are measured relative to a certain set of quality standards. For example, the overall quality of a given product is defined by a variety of characteristics: glossiness, thickness, scratches, foreign matter, dirt, cracks, etc. These can be measured quantitatively, and defect rates and remnant waste calculations can easily be made.

It is not easy, however, to determine at what point repair or tuning quality should be evaluated. Given a certain inherent anxiety, each step of maintenance work—such as disassembly and assembly—must be checked from beginning to end. If there is high dependence on outside subcontracting, there is no good way to do this except by reinforcing supervisory staff (which means an added indirect cost). But since there is a limit to the size of a capable supervisory staff, it is important to prevent a situation where supervisors who are less qualified than the subcontracting engineers for a particular job are simply bystanders.

Repair maintenance is checked by means of trial runs. If problems result, then the repair was the lowest quality. Even if the maintenance passes the test, we can expect problems to occur later. This means that the quality of repair and tuning work will have a profound effect on the equipment's life cycle, ultimately determining its reliability. It is also important to recognize that breakdowns occurring one or two months after such repair or tuning are hard to pinpoint as being due to defective maintenance or poor operation.

Thus, maintenance output is difficult to evaluate. While each worker's commitment should be to a "zero defect" campaign—error-free at every step—something beyond a representative set of values for quality or reliability should be sought. People sometimes say, "Since it was done by that project leader's unit, we can depend on the work." But what does this really mean? At the very least, it means that the

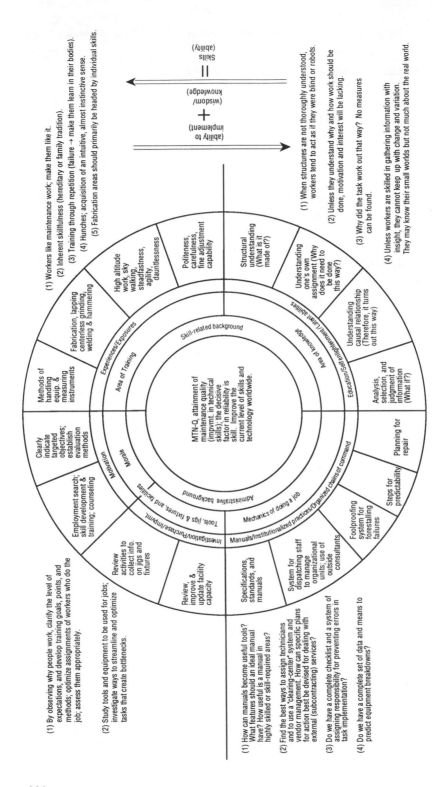

FIGURE 7.3 Improvement and Guarantee of Maintenance Quality (MTN-Q)

Skills (ability)

=

+ (ability to implement)

(wisdom/knowledge)

Skills (ability)

(1) Workers like maintenance work; make them like it.
(2) Inherent skillfulness (hereditary or family tradition).
(3) Training through repetition (failure → make them learn in their bodies).
(4) Hunches; acquisition of an intuitive, almost instinctive sense.
(5) Fabrication areas should primarily be headed by individual skills.

(1) When structures are not thoroughly understood, workers tend to act as if they were blind or robots.

(2) Unless they understand why and how work should be done, motivation and interest will be lacking.

(3) Why did the task work out that way? No measures can be found.

(4) Unless workers are skilled in gathering information with insight, they cannot keep up with change and variation. They may know their small worlds but not much about the real world.

High altitude work, sky walking, steadfastness, agility, dauntlessness

Politeness, carefulness, fine adjustment capability

Structural understanding (What is it made of?)

Understanding one's own assignment (Why does it need to be done this way?)

Understanding causal relationship (Therefore, it turns out this way)

Analysis, selection, and judgment of information (What if?)

Fabrication, lapping centerless grinding, welding & hammering

Methods of handling equip. & measuring instruments

Skill-related background

Experiences/Exposures

Area of Training

Development/Latent abilities

Area of knowledge

Education/Self

Planning for repair

Steps for predictability

Foolproofing system for forestalling failures

MTN-Q, attainment of maintenance quality (impvmt. in technical skills); the decisive factor in reliability is skill. Improve the current level of skills and technology worldwide.

Clearly indicate targeted objectives; establish evaluation methods

Motivation

Morale

Administrative background

Tools, jigs & fixtures, and facilities

Investigation/Purchase/Impvmt.

Employment search; skill development & training, counseling

Review activities to collect info. on jigs and fixtures

Review, improve & update facility capacity

Specifications, standards, and manuals

System for dispatching staff to manage organizational units; use of consultants

Mechanics of doing a job

Manuals/Institutionalized practices/Organized chains of command

(1) By observing why people work, clarify the level of expectations, and develop training goals, points, and methods; optimize assignments of workers who do the job; assess them appropriately.

(2) Study tools and equipment to be used for jobs; investigate ways to streamline and optimize tasks that create bottlenecks.

(1) How can manuals become useful tools? What features should an ideal manual have? How useful is a manual in highly skilled or skill-required areas?

(2) Find the best ways to assign technicians and to use a "clearing-center" system and vendor management. How can specific plans for action best be devised for dealing with external (subcontracting) services?

(3) Do we have a complete checklist and a system of assigning responsibility for preventing errors in task implementation?

(4) Do we have a complete set of data and means to predict equipment breakdowns?

290

repair and tuning quality are represented by the skills of that particular work unit.

Similarly, improvements in productivity through PM activities, as well as the extent of problems in tuning and adjustment type maintenance, are greatly affected by a worker's individual motives and personality. This also applies to production in general. Maintenance skills, which span a broad range of technical fields, need a number of years to attain respectable levels. In maintenance activities where a considerable depth of skills is also required, success depends on aggressive promotion of the skill training itself.

7.2.3 Verification of Techniques and Skills Through One-Point Lessons

In the past, skills were not thought of as "taught," but rather as "stolen" from one's superiors during apprenticeship. In either case, skill acquisition is not easy. In the current era of technological innovation, the attitude that anyone can acquire "indispensable" skills is outdated. Just keeping up today means technological regression. Thus, we need methods that provide more efficient and versatile technical skill education and training. In the final analysis, education and training are techniques that make it possible to transmit and verify communication skills to people who need them.

In PM education, one-point lessons are an especially important part of the curriculum. One-point lessons are a means of accumulating, transmitting, and verifying "first-hand" know-how from the shop floors. They range from work activities for equipment quality, safety, and operation, to those concerned with equipment function, structure, troubleshooting, and remedial action. One-point lessons should be written clearly. Visual aids, such as diagrams, should be used as much as possible to explain a single point, in conjunction with live demonstrations. The one-point lessons have two purposes: to "know how" (explaining how things should or should not be done) and to "know why" (explaining why things are or are not the way they should be). An accompanying table should have a column for checking whether a lesson has been assigned.

In addition to this verification mechanism, one-point lessons can be divided into those that transmit and those that verify knowledge. As in formal education, textbooks and exercise books can differ. Creating one-point lessons to verify acquired knowledge (e.g., tests) may be their main purpose. But, in terms of the aim of education—acquiring knowledge and skill—the one-point lesson carries learning by memorization further so that what is learned can be applied through transmission and verification.

	Transmission	*Verification*
Knowledge	Taught	Test
Skill	Learned, Tried	Can apply any time

One-point lessons should be taught whenever needed, and even seemingly trivial things in daily life should be compiled and documented. Through such practice, not only can technical skills be enhanced, but a mind-set and culture can be created in the workplace that is conducive to building the desired skills. Furthermore, categorizing and organizing the one-point lessons thus collected will also provide examples of skill training best suited to one's own work environment.

7.3 PROMOTING SKILL TRAINING

7.3.1 Management Leadership of Education and Training

As PM practices progress, there will be a need for training and education in more areas. Staff personnel usually design the curriculum and plan education and training programs on an annual basis. Although these plans may look good on paper, the staff is often too busy to implement them successfully. In some education programs, we saw a large number of attendees in the beginning, but attendance diminished as sessions continued until only a few participants were left; others were simply discontinued due to total audience attrition.

Dropping out may relate to upper or middle management's failure to recognize the need for such education and training programs. Unless management can motivate and encourage participation, clearly identifying what is expected to be gained from these programs, effective implementation will be difficult. Both upper and middle management should show interest and concern, visiting classrooms from time to time, and even taking part in teaching.

Another cause of stagnant education and training programs lies in the method of education itself. Usually, formal schooling is what comes to mind in talking about education, but this tends to be "knowledge-centered," and is often inadequate. Recent graduates from technical high schools with a diploma in mechanical engineering often cannot operate a single machine. Some say that they are uninterested in more education. Knowledge, undoubtedly, is important, but unless skills are acquired, the knowledge cannot be put into practice and will be of little use. In an environment of on-the-job training, 70 percent of the time should be dedicated to practical skill training.

It should be noted, too, that in skill training, some workers score poorly in the group education environment, but once in the actual shop environment, may become ace technicians. Conversely, some may score high marks in the educational environment, but will prove unproductive in a real job environment. Management should therefore provide an environment in which posteducation trainees can use their training and test their skills—by assigning them tasks to apply newly acquired skills in innovative maintenance, for example, and having them make presentations to their colleagues. Motivation of participants in training programs would thereby be enhanced, confidence would be gained by practicing new skills in a real environment, and peer recognition would raise self-esteem. This would enable trainees to advance even further and return the time and money spent in education many times over.

An atmosphere would also be created that encourages management to send successful trainees to further training programs. In this regard, the curriculum might need reorganization. Skill training calls for special environments and methods that make trainees feel enthusiastic about their training programs. A competitive system, such as a maintenance (skill training) rally, could be one such solution.

7.3.2 Student Awareness and Instructor Enthusiasm

In these days of perpetual engineering innovation, students need to be aware that engineering knowledge and technical skills must be continuously sharpened, lest they fall behind.

In this context, education and training can be seen as first serving the individual, and then the business or employer. If an engineer is singled out for a certain job in a given work environment, it is his or her responsibility to fully use his or her abilities and knowledge to accomplish the task. Thus, even if the engineer leaves the group, the education and training received remains his or hers. It is important that students fully understand this point, as well as the fact that education and training is always tested—whether in a classroom education or on the job.

The purpose of testing is not to grade individuals on a "one-time" basis, but to clarify areas still to be mastered so that they can be reinforced. Tests should be given repeatedly until the students pass them. The key to successful skill training lies in this cycle of patient, followup testing. This approach is particularly necessary because on-the-job training differs from formal schooling. Employees attending after-work education may experience problems at home and/or fatigue. Instructors must be committed and determined to continue educational efforts until the students reach the level of accomplishment expected by their superiors or leaders.

7.3.3 The Pair Method of Education

Education and training should be followed up through the pair method. Since PM activities are significantly affected by both knowledge and experiential expertise, the speed with which workers learn depends very much on the personalities of their leaders in the work environment. Workers might reach their leaders' levels relatively easily, but as long as they are exposed to nothing further, are not likely to surpass them. Thus, management should choose a number of thoroughly experienced technicians (regardless of rank), and have these experts serve as sponsors or mentors to the students. These relationships should be scheduled outside of regular work hours, and the pairing of mentors and students rotated from time to time.

7.3.4 Practical Education on the Job Site

Education should be closely tied to actual work in the job environment and study materials should integrate educational goals with job requirements.

In the beginning, educators often make the mistake of trying to train maintenance staff in reading engineering diagrams and drawing charts, although many maintenance crews rarely work this way. Even when repairs are subcontracted, trainers may resort to on-site explanations, or write out their bills cryptically. Taught this way, it is likely that the students may never even be able to draw a simple chart. If, on the other hand, the maintenance crew's station is equipped with many racks of bound drawings depicting the equipment for which they are responsible, and if the environment calls for the use of these drawings each time repair and maintenance is done, and if workers are accustomed to "red-pencil" problem areas on the drawings, then an environment exists in which the "how-to" of reading and drawing charts can be easily acquired as "living" education.

7.4 "HANDCRAFTED" MAINTENANCE SKILL TRAINING PROGRAM

Unlike formal schooling, in-house skill training follows a different pattern from lectures and pass-fail examinations. It is intended, rather, to train the talent needed in the business. That is, the objective is not to screen people, but to nurture them. These efforts cannot succeed unless creative thought is given to a methodology that incorporates the unique characteristics of each business.

One example of an in-house skill training program is shown in Figure 7.4, which pertains to a machine processing and assembly industry. It is divided into self-initiated and general maintenance skill training. In the former, each step accompanies a higher stage of advancement. The objective of each step is clearly defined, and is aimed at developing an instructor whose role forms the nucleus in self-initiated maintenance. Maintenance skill training, on the other hand, is aimed at more than training for a particular maintenance task, such as instrumentation or welding skills. It ranges from a course for inspection or production technicians to a course that trains workers for a particular job class, i.e., raising skill levels in electronics maintenance. Figure 7.4 also gives an example of the education curriculum, in this case, the intermediate course for self-initiated maintenance.

Since successful education depends on the way it is taught, the intermediate course goals are shown at each step. Specifically, the course follows the steps, teaches the knowledge, examines the level of understanding, trains through skill education, verifies through "devil's advocate" diagnostics, applies what was learned in a "live" environment, and finally, presents a self-critique and review. This systematic approach shows the students' transition from "I don't know" to "I know now"; from "I can't do it" to "I can do it"; and from "I won't do it" to "I did it." This is a revolutionary process of changing people.

In-house training programs—including textbooks, materials, and evaluation methods—should be well suited to the needs and characteristics of each company. Textbooks, rather than being comprised of basic knowledge or past problems, should be made from scratch, no matter how thin they may be. Tools for skill training—cutaway models and simulation devices—are fine, but unless they too are "homemade," will be difficult to use. Evaluation should focus on each spe-

Self-initiated intermediate course curriculum

Steps	Subjects	Time	Contents	Materials	Offered by:
Getting to know equipment (knowledge)	Electrical	4 hrs.	• Sequencing knowledge • Overall safety knowledge	• Textbooks • Cut-out models • Skill tests	I. Johnson
	Mechanical	4	• Knowledge about air	• Textbooks • Cut-out models • Skill tests	Tassari Silver
Trying to make equipment (skills)	Electrical hands-on	8	• Sequential assembly	• Relays, limiters, timers	Hughes Kenyon
	Mechanical hands-on	8	• Assembly of air circuits	• Cylinders • Valves, hoses	R. Johnson Saito
Devil's advocate method of diagnostics & repair (verify)	Equipment diagnostics	8	• Diagnostic knowledge • Anticipatory breakdown detection	• Textbooks • Simulation equipment	Logan Gillespie Cato
	Line diagnostics	8	• Servicing equipment failure spots		Newcomb Drezinsky Howard
Knowledge & skills	Press molds	8	• Functions of molds • Maint. & adjustment of molds	• Textbooks	Short Erhardt Higgins
Practical experience	On-site training	16	• Practical experience in maint. & impvmnts. • Presentation of experience • Certification	• Maint. people or instructors should be mentors	Participants: plant mgr., directors, & mgrs.

Goals

People who can do minor repairs

Do not know equipment ⇒ Training through group education → Getting to know equipment → Can do minor repairs & improvements → On-site on-the-job training (OJT) ⇒ Can make equipment

Education/ training course YY/MM	Target head count	'84				'85												'86		
		9	10	11	12	1	2	3	4	5	6	7	8	9	10	11	12	1	2	3
Self-initiated maintenance skills — Self-initiated maintenance instructor cultivation (10 hrs.)	50	On-the-job training (6 times, already completed)																		
Self-initiated in-house maintenance (beginner) (10 hrs.)	400																			
Self-initiated maintenance (intermediate) (48 hrs.)	160			11		8	8	11	16	15	11	8	10	10						
Self-initiated maintenance (advanced) (3 mos.)																				
Mold handling (8 hrs.)																				
Maintenance skills — Measuring skill (3 hrs.)																				
Mechanical foundations (maintenance work) (3 hrs.)																				
Mechanical fabrication (2 hrs.)																				
Electronics maintenance skill																				
Welding skill (4 hrs.)																				
Diagnostic skill																				
Maintenance "rally" (3 mos.) (Pneumatic, hydraulic, mechanical, electrical, material, & planning)																				
Methods study session																				

FIGURE 7.4 Maintenance Skill Training Plan

cific item, using written and actual skill tests. The maintenance crew members who will serve as the skill training instructors do not usually have a chance to make a speech before a sizable audience. Thus, teaching and creating textbooks and instructional materials will be their best education. While this self-education process is taking place, enthusiasm about teaching can be expected.

8

Effects of Preventive Maintenance and Maintenance Evaluation

8.1 SIGNIFICANCE OF MEASURING THE EFFECTS

Problems in measuring the effectiveness of PM activities must be dealt with to advance PM. This issue is complicated by the meaning of the term "effect," which has multiple connotations. In this context, it might be helpful replace the word "effect" with "evaluation" or "accomplishment." For example, the effect of PM activities can be expressed in the 5S level, its effect and value, and evaluation and effects of the educational measures taken. It should also be noted that the effect may be quantitative or qualitative.

When a plant adopts a PM effect-finding system, unless the real meaning of measuring effects is thoroughly understood, the result will be mere data gathering, an enormous waste of time.

Various indexes for measuring the effect of PM are similar to indexes of defect rate statistics, lot-sizing statistics, cost-burden ratios, accident rate frequency, or the rate of missed delivery dates. They deal with events occurring in a certain time period that are quantitatively calculated according to fixed categories. In other words, they are indicators of efforts made by each department over a given period of time. Stated simply, they are only "scorecards."

Another way to think of them is as students' school records. They assess work done in a given period, and fall into the category of management data. As management data, their contents should be analyzed and the most important findings used to determine the next steps to be taken in cooperation with one's superiors. These records can also inspire workers to "do better next month," and, as such, can be considered indexes of psychological stimulation.

Management data serve two purposes. One is to enable personnel outside a given department (particularly supervisors or upper management) to see if problems can be solved by the personnel directly involved, and, if so, what measures should be taken thereafter. The second purpose is to allow supervisors and upper management to demonstrate their interest in the data, by either commending or reprimanding workers as appropriate. This is crucial if the data are to be of management value. If they are not used in this way, their role as psychological stimuli, inspiring personnel to "do better next month," will disappear.

If supervisors excessively criticize and accuse employees for shortcomings and nag them about details, the effects could seriously backfire: data with excuses and explanations will simply increase in volume. In extreme cases, an employee may succeed in "selling" excuses to his or her supervisors and breathe a sigh of relief. At this point, the situation is hardly amusing. On the other hand, if supervisors are indifferent to management data, then employees do not feel inspired to say to themselves, "Let's do better next month." Thus, management data will either "live or die" depending on management's capabilities. Although rather delicate, these data are still an important management tool.

Management data differ substantially from engineering data. Workers with a sense of responsibility are well aware before the management data are tallied whether the current month's output is falling, and if so, for what reasons—accidents, problems, etc. Genuinely qualified shop floor supervisors can predict how their section is performing relative to the previous month. In fact, this is a criterion of their professionalism.

In many cases, many kinds of data abound in shop floor activities. Unless the meanings and the characteristics of each type of data are well understood, we tend to get misled into thinking that abundance itself of management data and engineering data means that the system is well managed. It should be noted that for all the abundance of data, much may not necessarily be useful, if not obstacles.

The method of measuring the effect of PM must fit in with the plant's management level. This may stimulate thought about the acute need to investigate the organization of the unit actually doing the work.

Many elegant evaluation methods developed by staff personnel do not fulfill their intentions. Measuring the effect of maintenance starts with clearly indicating the annual goals set by upper and middle management and targeting current efforts. A measurement method and a managerial unit that fit those objectives should be worked out through joint efforts of management and staff.

Unlike engineering analysis data, management data define the relationships between causes and effects. When a mistake is made, its cause must be sought so that remedial action can be taken immediately, and repetition of the error avoided. For example, this means that the breakdown rate for the current month is stated, and the losses derived from production stoppages are noted as well. These figures may be meaningful as management data, but are totally useless as engineering analysis data. What must not be forgotten, however, is that engineering analysis data cannot give impetus to managerial units. In other words, to provide "stimulation," there are more appropriate ways to gather data.

8.2 ENHANCING EFFECTIVENESS THROUGH THE PM RALLY

8.2.1 Improving Management Ability of Line Managers

Organization is the key word to describe contemporary business functions, an organizational concept shown in Figure 8.1. Recently, businesses have been using organizational systems conducive to goal-

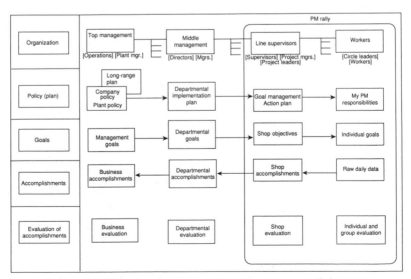

FIGURE 8.1 Links Between Goals and Accomplishments

oriented management. Through these systems, corporate strategies are hammered out, as well as plant goals and target values. These filter down the echelons step by step, from departments to workplaces, through each managerial level. At the top, high-level corporate strategies or slogans are posted, but we often wonder whether they truly trickle down to each worker, where they should serve as the driving force of business conduct.

There is a children's game called "telephone." Several people sit in a line or circle, and, starting at one end, a message is whispered to the next and passed along in a chain. Usually, the last person who hears the message says something entirely different from what was said at the start. This highlights one aspect of human communication. If a corporate or plant slogan is prominently posted, everyone will see it. But unless it is learned by heart and recited, it will not affect how business is conducted. As an organization becomes multilayered, communication will tend to become increasingly inaccurate.

Goal-oriented management tends to set fine goals, but may lack the means to attain them, or may neglect them. The fact of the matter is that results are achieved in the workplace itself, and workplace and departmental evaluations, or assessments of managerial accomplishments, depend on the accumulation of seemingly trivial and ordinary results that occur where the work is done.

What this simply means is that results are achieved through the activities of each individual shop floor worker, and outcomes are determined by how thoroughly work is done.

The recent generation of shop floor line supervisors is good at managing people, but few are both skilled technicians and leaders. We share the concerns of critics who question the acceptability of this trend. Line supervisors give instructions, advice, and guidance to shop floor workers, directing the group as a whole. Strengthening their capabilities thus means strengthening group work habits. Managers produce results only through their work units.

In many cases, however, line supervisors may simply receive one-sided instructions and goals from their managers. This relationship cannot be expected to foster a strong work environment. The key to improving line supervision is to develop effective managerial and leadership skills, the essence of which lies in improving line supervisors' skills in planning, management, leadership, and implementation. This also means creating a disciplined work environment that can accomplish some tough objectives.

Have sufficient efforts been made to strengthen the line supervisors—the key personnel? What methods are used to train them? What areas should be emphasized? What should be done on a day-to-day basis to progress in this direction? Managers must not assign unilateral goals. Rather, they should try to improve the quality of line supervisors and workers, and provide guidance to make their workplace individually fulfilling.

As one method of strengthening line supervisors, we are introducing here "a goal management plan table" or "PM rally." We also propose the idea of "my TPM responsibility" for each individual worker. Through such methods, we expect that an organization or system can be devised that is very responsive to external and internal stimuli, integrating all levels from upper management and department heads down to individual workers.

8.2.2 Goal Management Plan Table

GOALS

This table is a tool with which line supervisors can establish a specific set of goals, and work on planning, executing, and evaluating them. Using this tool, line supervisors can improve their planning and managerial abilities and learn the art of leadership, while attending to the details of daily, weekly, and monthly management cycles.

Line supervisors trained in this way can not only improve their work environments, but can also see how to help their subordinates grow professionally and become capable managers themselves, interested in understanding each individual line worker. The planning table's format is shown in Figure 8.2.

Timely entry of remarks should be made in the table, at no more than weekly intervals. Supervisors should help define its main topics on a monthly basis, and a critical review should be done monthly to plan for the month ahead. It is crucial that goals be clearly quantified. Since there will be no more than twelve tables per year, they should be kept for future reference.

Every two or three months, the table should be used to make presentations to supervisors or peers, to point out problems, and to assist in cooperative planning efforts. Plans are much more than a calendar of schedules, and it is important that action plans be written into the table.

8.2.3 "My PM Responsibilities"

It is easy to say that each worker should accomplish the goals set. To do so is another matter. People do not intend to break promises, but

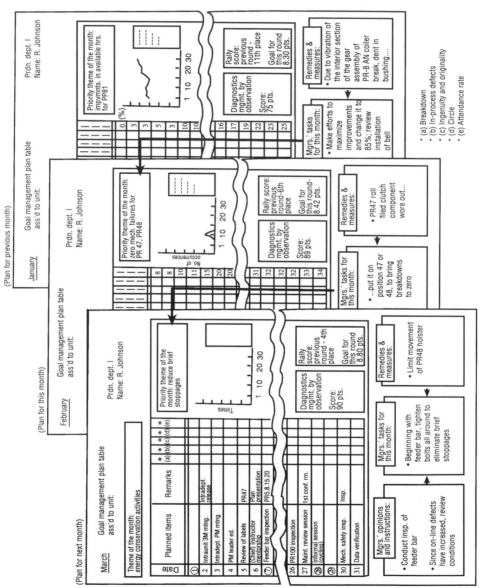

FIGURE 8.2 Goal Management Plan Table

are forgetful, find tasks cumbersome or awkward, and overlook things. This is why foolproofing, equipment augmentation, and training are necessary.

The participation of each worker is needed to reach management goals. Thus, verbal initiatives should be taken and acted upon. The key to success lies in habitually remembering to act upon what is initiated. Implicit in the concept of "my TPM responsibilities" is a reminder to all workers that each individual accepts responsibility for and participates in TPM. Although the managers are responsible for the final outcome, workers need to accept the philosophy that each is responsible for his or her own behavior. The responsibilities must be written down by each worker, under supervision, and should be revised from time to time. What needs to be done should be clearly defined, and should be posted on a bulletin board or written on cards that each worker can carry. Their contents must be learned by heart, to be recited from memory, and so will only occasionally need to be checked by supervisors. Items can be divided into categories, such as quality, production, maintenance, safety, and the 5S's.

Whether defined as a goal management plan, "my assigned responsibility," or "my pledge," the common aim is to reach the same objectives as the principles underlying creation of a desirable work environment in the 5S tradition. These are:

1. Developing a sense of individual responsibility for assigned tasks;
2. Educating workers to do these tasks;
3. Teaching strong mathematics skills.

8.2.4 A Competitive System—The PM Rally

1. RALLY GOALS

The PM rally is competitive, but as discussed here has a different goal than competition: that relates to the current state of PM activities. The effectiveness of conventional PM activities—defined as equipment maintenance—is judged by the degree of equipment improvement or by a reduced breakdown rate. This is true even though it is often unclear how PM activities are related to an improved work environment. This situation pertains to PM when it is considered as just a type of maintenance activity only minimally oriented toward total companywide commitment. The PM rally seeks to systematize PM and strengthen its approach to equipment-oriented management.

In this section, we shall deal with PM target values as if they were on the same plane as the business goals of any given firm. This will enable us to link line supervisors conceptually to the business organization. Thus, within each work environment PM activities and production activities are no longer separate. The point has often been made that PM and production activities can be the same. This point need no longer be proselytized or pushed, since line supervisors can introduce PM while recognizing where their activities are headed. Furthermore, line supervisors can compete with their peers in the same sets of categories to be evaluated. Through this competition, their work environments will be improved.

Target values have traditionally been handed down by upper management and regarded as if they should be accomplished without question, and, if not, then middle management and line supervisors would be questioned. Since individual line supervisors can influence results, however, target values are meaningless unless they are accepted on the shop floor itself. Therefore, by forming a group made up of workers and their line supervisor, the efforts of any key individual would immediately show up and could be evaluated. When such evaluation groups are set up, through joint efforts within each of them, mutual enlightenment and training can be expected as each group "gels" as a team.

Managers, supervisors, and staff will be able to see improvements in these groups, expressed in higher scores, and will be encouraged to use them to reach business goals. The roles of managers and staff will inevitably become clearer as these groups continue to function.

2. THE PM RALLY

Figure 8.3 shows how a PM rally works at company "M."

The main tool of the PM rally is the goal management plan table. It is implemented on the basis of departmental and sectional goals, and plan tables are submitted by each work group. Results are judged on the basis of the evaluation standards. Through this process, an order of rank is established among work groups, while the scores for evaluating accomplishment levels are displayed for each item and for overall performance. Announcing the ranking order fosters a competitive spirit. If a work group remains at the bottom for several months, management and staff may be considered as lazy. However, positive reinforcement rather than criticism should be used to inspire the group to make efforts to improve.

If evaluation results rise, it means that the work environment and business are improving. It is important to identify which evaluation items showed poor results and which need improvement, so that plans

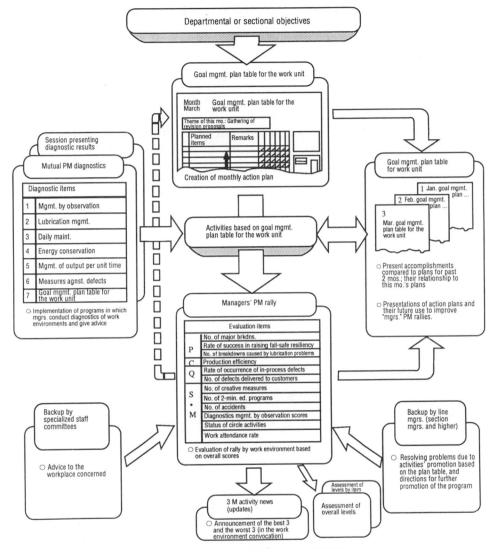

FIGURE 8.3 Mechanics of the PM Rally

can be made. Several work groups should present plan tables at the same time, so that each can learn from the others about how to make and implement plans.

Since supervisors, staff, and special committee organizations are all graded at these PM rallies, before the rallies are held, consideration should be given to what kind of support the work groups need in order to improve. The PM rally is aimed at improving managerial skills through line supervisors, and efforts to achieve this should not be stymied in the middle. Thus, promotion of this type of program may require building an infrastructure to improve the work groups' overall attitudes.

3. DEFINING AND SCORING EVALUATION ITEMS

Items to be evaluated should match PM and production activities in the work environment. They should be chosen from among

- P (production),
- Q (quality),
- C (cost),
- D (delivery),
- S (safety and pollution prevention), and
- M (morale).
- If only measurable, quantifiable targets are selected, the evaluation will score only results. It is important, therefore, to include in the evaluation the degree of effort made in doing work. Evaluating the level of effort made can compensate for possible poor results.

This consideration is based on the following. In the business world, scores must be measured in terms of accomplishments. But there are differences in levels of difficulty among work environments, as well as unexpected or extenuating circumstances. A results-only evaluation would not be able to judge whether the work environment is indeed improving or is heading in the right direction. A positive, forward-looking attitude is crucial for line supervisors and workers. In the long run, this should yield good grades and scores, but, in a sense, these must be understood as independent of broader goals. Thus, evaluation of results accomplished must be combined with evaluation of the level of effort put into the work.

Target items should be defined with maximum clarity so that the line supervisors and workers can understand them at a glance. For this purpose, it is preferable that absolute values be used. Items are also needed that can be used to compare work groups, sections, or departments.

Evaluation items should be weighted differently from time to time. For example, if the following is a PM month, then "breakdowns" would be weighted most heavily. A weighted point system may work well in this regard.

- Evaluation standards should be fair and unbiased, to enable anyone to judge the weaknesses of a work group and define how to correct them.
- The number of items to be evaluated poses a difficult question. Items should be simple at first, and gradually increase in number and complexity.
- The evaluation of a PM rally is a mirror that reflects the reality of the work environment.

The evaluation items and role of line supervisors are displayed in chart form in a matrix on monthly scorecards. For each evaluation item, comments and advice are given by the special committee that handles the item, or by the overall corporate organization. For example, breakdowns will be discussed by the MTBF special analysis committee, and accidents by the safety and public health committee. Moreover, line supervisors will be advised by their superiors.

8.3 CHOOSING ITEMS FOR MEASUREMENT AND EVALUATION OF IMPACT AND EFFECT

Many factors are included in evaluation items. In this case, the level of management sophistication should be determined. This will vary depending on the scale of the business. Items managed at each plant should be organized, and those to be evaluated should be selected on the basis of management items that are common to several plants. However, characteristics of manufactured products and differences among manufacturing processes may determine what management points should be emphasized. These should be resolved during the goal-setting stage, after careful examination of common management items.

At the beginning of this book the goal of PM activities was defined as enhancing equipment's functioning in response to plant management's output targets: i.e., production volume (P), quality (Q), cost (C), delivery (D), and safety, health, and pollution prevention (S). It follows that in measuring PM's effect, equipment output must be viewed from the standpoint of each evaluation item. Furthermore, not only the output, but the input into the equipment should be reviewed as well. For example, the costs of equipment sustenance, production improvement, maintenance improvement, and other capital outlays must be reviewed as evaluation items.

Special attention should be given to the way PM activities are promoted, a process that cannot be quantitatively measured. For example, what is the lowest corporate level to which the company's PM goals have reached? Even in the absence of noticeable or tangible effects, the question arises as to how to evaluate the process whereby efforts are being made. This is an extremely important question. The lower down we go in the corporate ladder, the stronger our feeling is that efforts should be more carefully evaluated. This is because the ultimate purpose is to inspire people with the will to say, "let's work harder," and to give them encouragement. This is the essence of measuring impact and effect.

Readers are referred to other authoritative publications with regard to kinds of evaluation items. Some typical ones are shown in Tables 8.1 and 8.2.

TABLE 8.1 Examples of Measuring the Effects of PM

Managerial aspects	Maintenance aspects
① Sales accomplishments ② Operating profits ③ Value-added productivity ④ Return on investments	① Assessment of maintenance standards level ② Rate of accomplishments of planned maintenance ③ MTBF - MTTR ④ Number of cases of corrective maintenance ⑤ Amount of spare parts and stock inventory - inventory turnaround

Plant management (main index)		Plant management (subsidiary indices)
① Rate of equipment utilization for output (Rate of equipment effectiveness) ② Work productivity ③ Sudden breakdowns and fail-safe resiliency ④ Number of brief stoppages	P	① Hours when equipment was not utilized ② Number of major breakdowns ③ Breakdowns caused by lubrication problems ④ Assessment of QAL level ⑤ Number of quality components ⑥ Rework ⑦ Yield ⑧ Electricity and lubricant consumption ⑨ Rate of attainment of budgetary goals for maintenance ⑩ Number of transitions to in-house manufacture of equipment ⑪ Number of head counts saved ⑫ Number of vehicles dispatched in abnormal situations ⑬ Accidents resulting in plant shutdowns, accidents resulting in effects short of shutdowns, accidents requiring some minor patchup ⑭ Skill training ⑮ Number of times one-point lessons practiced
⑤ Market warranty disbursement amount ⑥ Occurrences of vendor defects ⑦ Rate of in-process defects	Q	
⑧ Monetary valuation of streamlining and realignments ⑨ Monetary valuation of material and energy conservation ⑩ Inventory cost	C	
⑪ Rate of on-time deliveries	D	
⑫ Number of accidents ⑬ Success in meeting antipollution regulations	S	
⑭ Assessment of circle activities ⑮ Rate of confirmation of the 5S's ⑯ Number of original and ingenious suggestions	M	

TABLE 8.2 Method of Measuring Effects of Maintenance (based on DuPont Method)

Basic factors	Evaluation factors	Formulae
1. Planning characteristics	1-1 Planned activities	$\dfrac{\text{Hours of planned activities actually worked}}{\text{Hours actually worked}} \times 100$
	1-2 Unplanned activities	$\dfrac{\text{Hours of unplanned activities actually worked}}{\text{Hours actually worked}} \times 100$
	1-3 Labor effectiveness	$\dfrac{\text{Hours actually worked}}{\text{Hours of attendance}} \times 100$
	1-4 Overtime labor	$\dfrac{\text{Overtime hours actually worked}}{\text{Hours of attendance}} \times 100$
2. Amount of work	2-1 Current volume of work	$\dfrac{\text{Weekly average hours of work}}{\text{Weekly capacity hours}} \times 100$
	2-2 Total volume of work	$\dfrac{\text{Month-end total hours of work}}{\text{Weekly capacity hours}} \times 100$
	2-3 Preventive maintenance	$\dfrac{\text{Hours worked on PM}}{\text{Hours actually worked}} \times 100$
	2-4 General maintenance	$\dfrac{\text{Hours actually worked on general maintenance}}{\text{Hours actually worked}} \times 100$
3. Costs	3-1 Repair costs	$\dfrac{\text{Repair cost for this month (\%)}}{\text{Previous year's monthly average repair cost (\%)}} \times 100$
	3-2 PM costs	$\dfrac{\text{This month's PM cost (\$1,000/Unit)}}{\text{Previous year's monthly average PM cost (\$1,000/Unit)}} \times 100$
	3-3 Direct cost	$\dfrac{\text{Direct repair cost (\$1,000)}}{\text{Total equipment repair cost \ (\$1,000)}} \times 100$
	3-4 Indirect cost	$\dfrac{\text{Indirect repair cost (\$1,000)}}{\text{Total equipment repair cost \ (\$1,000)}} \times 100$
4. Productivity	4-1 Rate of productive maintenance	$\dfrac{\text{Maintenance cost (\$1,000)}}{\text{Production output (\$1,000)}} \times 100$
	4-2 Production stoppage losses	$\dfrac{\text{Stoppage hours}}{\text{Equipment available hours}} \times 100$
	4-3 Rate of cost of production	$\dfrac{\text{Total departmental costs}}{\text{Output (\$1,000)}} \times 100$
	4-4 Rate of actual operation time	$\dfrac{\text{Standard work hours (in minutes)}}{\text{Hours actually worked (in minutes)}} \times 100$

Bibliography

Nakajima, Seiichi (ed). *Opereita no tameno Setsubi Hozen Kosu* (Equipment Maintenance Course for Operators). Japan Management Association, 1981.

———. *Setsubi Hozen Jissen Kiso Kosu*. (Equipment Maintenance Practitioner: Basic Course). Japan Management Association, 1979.

Osada, Takashi. *Setsubi Gijutsu Engineer*. (Equipment Technology Engineer). Institute of Productive Maintenance Technology, 1983.

Takahashi, Yoshikazu. *Seisan Hozen Suishin Manual*. (Manual of Productive Maintenance Advancement). Japan Institute of Plant Maintenance, 1975.

Case Study in TPM Activities on the following companies:

Aichi Steel Works Co. Ltd.

Aishin A.W. Co. Ltd.

Aishin Seiki Co. Ltd.

Aishin Takaoka Co. Ltd.

Asahi Chemical Industry Co. Ltd.

Daiken Trade & Industry Co. Ltd.

The Furukawa Electric Co; Ltd.

Futaba Industrial Co. Ltd.

Maruyasu Industries Co. Ltd.

Nippon Zeon Co. Ltd.

Tokai Rika Co. Ltd.

Toyoda Iron Works Co. Ltd.

Yamaha Motor Co. Ltd.

Note: All works are in Japanese.

Index

Abnormal symptoms, 195
Abrupt breakdowns, 55-57
Absenteeism, 114
Accidents, 17, 117
  causes of, 119-22, 130
  prevention by 5S activities, 125, 130
  see also Safety, environment, and pollu-
    tion factors (SEP)
Action Then Theory (Toyoda), 14
Agricultural equipment, 4
Annual maintenance
  calendar, 175-76
  plan, 178-79
Antipollution technology, 2
Armed forces, U.S., and division of labor
  concept, 8-9
Assembly
  and block replacement method, 216
  just-in-time system, 14-15, 113-14
  and SEP factors, 115
  stoppages, 67-70, 119-22
  time-cycle setting, 50-54
Automation
  and broadened maintenance technology
    areas, 13
  foolproofing device, 92
  increased production line productivity, 51-
    52
  for inspection equipment, 86-87
  and production cost, 106
  robot uses, 3, 4, 30-31
  see also Computers
Automotive industry, 4
  see also Toyota Motors
Availability, 157

Behavioral science, 165-68
Best Business Environment with PM in Practice Award, 95
Block replacement method, 215-16
Bolts. *See* Tightening of equipment
Bottlenecks, 51-54, 164
Breakdowns
  analysis and remedies, 62, 64, 117-18,
    156, 192-94, 209-12
  and clearly defined responsibilities, 186
  crisis response, 166
  critical component analysis, 64-66
  and excessive spare parts inventory, 224
  and 5S activities, 126
  FMEA analysis, 271
  generating work reductions, 162
  interval irregularity, 205-6
  long-term, countermeasures, 213-14
  maintenance, 177, 178
  and missed deliveries, 114
  MTBF Analysis Table, 72-80, 85
  MTTR Analysis, 208-9
  patterns, 277-78
  preventive measures, 60-66, 126, 166,
    192-94, 230-37
  production workers' handling of, 229
  production workers' prevention of, 60-66
  reasons for, 47, 48, 55-57, 95, 97, 192,
    194, 205, 231, 232, 234
  recognition of occurrence, 210
  reduction in, 106
  repair time analysis, 209-12
  stresses-strengths margin, 192-94
  symptoms, 194-95
  types and resultant losses, 55-59

Breakdown (*continued*)
    and use of operations manual, 236
    zero, 44-50, 93-94
    see also Failure; MTBF Analysis; Stop-
        pages
Brief stoppages. *See* Short stoppages

CAD. *See* Computer-Aided Design
CAE. *See* Computer-Aided Engineering
Calendar, maintenance, 170-76, 185
CAM. *See* Computer-Aided Manufacturing
Catastrophes, 115
CC. *See* Cost control
CCA. *See* Critical component analysis
CD. *See* Cost down
Certification system, 246
Cleaning (*seiso*)
    as 5S activity, 32, 122, 129-30, 131, 135-
        40
    in self-initiated maintenance program,
        230, 231-32, 237
Color scheme, workplace, 129-30
Communication, 210, 302
Competitiveness, 251-53, 305
Computer-Aided Design (CAD), 280
Computer-Aided Engineering (CAE), 280
Computer-Aided Manufacturing (CAM), 280
Computers, 3, 48-49, 157, 235
Conservation. *See* Energy conservation tech-
    nology
Construction machinery, 4
Corrective maintenance, 177, 178
Cost control (CC), 23
Cost down (CD), engineering technology de-
    sign for, 262, 263, 264-67, 268, 269
Costs
    differentiation between product and equip-
        ment, 257-59
    equipment relationship, 22-23, 105-13
    evaluation of total maintenance, 187
    initial investment, 265
    life-cycle concept, 22-23, 76, 248-49,
        259, 262, 263
    MTBF Analysis Table for maintenance,
        109-10
    of PM, 27
    product, 257-59
    reduction of, 108-13, 126, 152
    spare parts management, 216, 223-26
    sustenance, 265
Cost tree, 266-67
CP. *See* Process capability
Critical component analysis (CCA), 64-66
Critical equipment, 61-62, 64-66
Cycles, maintenance, 198, 199

Daily equipment maintenance, 114, 237-42
Data, 71, 74-75, 300-301

Database, for building technical know-how,
    278, 280
Defective products, 114
    see also Quality control; Zero defect
Delivery date, equipment relationship analy-
    sis, 113-14
Deming Award, 95
Department of Defense, U.S., 22-23
    see also Armed Forces, U.S.
Design
    cost down awareness, 264-67, 268, 269
    of equipment engineering project, 260-64
    for maintenance prevention, 267-72
    of maintenance system, 152-57
    reliability/maintainability, 74-75, 256-57,
        271
Deterioration
    breakdowns from, 55-57, 95, 97, 194,
        278
    equipment inspection for, 189-92, 196
    measurement of, 191, 241
    prevention of, 230-37
    of spare parts, 219
Developing countries, renovated equipment
    in, 47
Diagnosis
    of 5S activities, 142-45
    of plant management standards, 43-145
    of PM system, 38-41
    for predictive maintenance, 189-201
    technologies, 75
    time and accuracy, 210-11
Diagrams and drawings, 76, 87
Dirtiness. *See* Cleaning
Disassembly, 194, 211
Discipline (*shitsuke*), 122, 130, 131
Division of labor, 227
    in Japan, 8-9
    for maintenance assignments, 73
    and skill inventory, 12-13
    in U.S. armed forces, 8-9
Drawings. *See* Diagrams and drawings
Dry-run times, 58
DuPont method, 187

Educational materials. *See* Manuals
Education. *See* Training within industry
Efficiency
    equipment analysis, 157-62
    equipment improvement, 22-23
    and equipment-production volume, 44-50
    of equipment repair system, 108-9
    and 5S activities, 125
    of maintenance system and work crew,
        154-55, 201-4
    and output improvement, 50-60
Electrical wiring, 236

Electronics, 2-3, 235
see also Computers
Employees. *See* Work force
Energy conservation technology, Japanese, 2-3
Engineers. *See* Equipment engineering technology; Industrial engineering; Value engineering
Environment. *See* Safety, environment, and pollution factors (SEP)
Equipment
  "aftercare", 253
  block replacement method, 215-16
  brief stoppages, causes and remedies, 67-70
  changeover and idle time, 49
  cleaning, 230, 231-32
  cost-efficiency improvement, 22-23, 105-13
  costs, 257-59
  critical, 61-62
  critical component analysis, 64-66
  delivery date relationship, 113-14
  deterioration, 57, 189-92, 194, 196
  diagnostic technologies in, 75
  economical stoppage plan, determination of, 48-49
  effective maximum use of, 285
  effectiveness ratio, 50
  efficiency analysis, 157-62
  evaluation cycles, 109-10
  failure analysis, 165-68, 201
  failure mechanisms and modes, 192-94
  failure prevention, 158-62
  flow of materials, 122, 127-128
  foolproofing devices, 92-93
  in-house creation of, 253, 254-255, *see also* Equipment engineering technology
  inspection of, daily, 237-42
  investment in, 46-48
  life cycle, 247, 248, 277
  life cycle costing, 22-23, 76, 248-49, 259, 262, 263, 265
  life cycle definition, 249-51
  life span, 49, 191
  lubrication of, 156, 231, 232-34
  maintainability, 74-75, 256-57
  maintainability improvement, 163-65, 201-14
  maintenance costs, 258-59
  maintenance costs ratio, 107
  for maintenance crew, 149
  maintenance methods, 176-78
  maintenance records, 70-85
  maintenance-related skills, 191, 285-97
  as major production means, 1-2
  manuals, 75-76, 259
  MTBF Analysis, 72-85, 109-10, 262

Equipment (*continued*)
  obsolescence cycle, 251
  operation of, correct, 236
  and operations characteristics, 55
  and plant security evaluation, 115-22
  PM use of stoppage time, 61
  precision-product quality relationship, 5, 89-91
  preventive maintenance, 2-5, 11, 157, 172-73
  prioritized, 62, 63, 168
  process capability of, 275, 276
  and quality issues, 75, 85-113, 231-32
  reliability, 74-75, 256-57, 271
  reliability improvement, 162-63
  renovated, 47
  repair management system and methods, 108-9
  replacement and modification of, 109
  selection in terms of PM management, 61-62
  self-initiated maintenance, 227-46
  SEP relationship, 115
  shop floor problems, 229-30
  shopping for, 251-53
  stresses and strengths, 192-94
  temperature control, 235-36
  tightening of parts, 231, 234-35
  utilization of existing during slow-growth periods, 7-8
  utilization rate, 44-50
  worker attachment to, *see* 5S activities
  worker familiarity with, 11, 122, 242-44
  worker training in use of, 285-97
  see also Breakdowns; Equipment-oriented management; Inspection; Parts; Production; Repairs; Spare parts
Equipment engineering technology, 247-84, 285
  activity-task zones, 259-60
  costs and, 257-59, 264-67, 268, 269
  data collection, 278, 280
  data and know-how, 256-57
  foundation for, 253-54
  goal setting, 263, 264
  importance of, 247-59
  improving company's skills, 251-53
  initial problems, 254-55
  and initial material flow management, 272-78
  know-how development, 278-80
  and maintenance prevention, 267-72
  project implementation, 260-64
  project summary and evaluation, 272
  promotion of workers' training in, 292-95
  task organization, 281-84
Equipment-oriented management
  during slow-growth periods, 5-8

Equipment-oriented management (cont.)
    importance of, 1-19
    and maintenance prevention, 247, 262,
        263, 267, 274-78
    and PM rally, 305-8
    and worker education and training, 7-8,
        285-97
Errors. *See* Human errors
European Conference on Maintenance, 113
Evaluation
    of 5S activities, 144-45
    defining and scoring items, 308-9
    of maintenance plans, 186-87
    of PM activities, 299-311
Exhaust restriction, Japanese, 2
Expertise, maintenance department, 149

Fabrication time, spare parts, 211
Factory
    cleanliness, 32, 129-30, 135-40
    equipment maintenance costs ratio, 107
    equipment as major production means, 1-2
    PM and plant management methods, 23-31
    rational organization of facilities, 32
    security systems, 115-19
    slow-growth periods, 5-8
    see also Equipment; Plant management;
        Production; Safety, Environment, and
        Pollution factors (SEP)
Failure
    analysis, 165-68, 201
    mechanisms and modes, 192-94, 271
    prevention, 158-62
    see also Breakdowns; MTBF Analysis;
        MTTR Analysis; Stoppages
Failure mode and effects analysis. *See*
    FMEA analysis
Fault Tree Analysis (FTA), 88
Fires, equipment, 236
5M Management, concept of, 27-29
5S activities, 32, 122-45, 156, 305
    concepts of, 123-24
    diagnosis and evaluation, 142-45
    goals of, 124-26, 130-31
    idea of, 122-23
    and management by observation tools,
        131-45
    manuals, 141-42
    meaning of, 127-31
    promotion of, 140-41
    slogan examples, 124
    three pillars supporting, 130-31
Fixed order point methods, 220, 222
Fixed-period ordering method, 222
Flexible manufacturing system (FMS), 3, 5
Flex-time work hours, 60
Flow of materials, 122, 127-28

FMEA analysis, 271
FMS. *See* Flexible manufacturing system, 3
Foolproofing devices, 92-93
Formulas, for rate of equipment effective-
    ness, 50
FTA. *See* Fault Tree Analysis

Garbage. *See* Cleaning
General Motors, 30
GNP. *See* Gross National Product
Goals
    of equipment engineering, 263, 264
    of 5S activities, 124-26, 130-31
    of maintenance records, 71-72
    management, 29, 31, 32-34, 302-5
    methods for setting specific PM, 32-34
    of MTBF Analysis, 73-76
    of PM system, 147-48
    of predictive maintenance, 190, 199-200
    of spare parts management, 215-17
    total vs. departmental productive mainte-
        nance, 31
    of TPM, 31-34
    zero WIP parts inventory, 16
Gross National Product (GNP), Japanese, 1,
    2
Groups, to promote productive maintenance,
    37-38

Handbooks. *See* Manuals
Hardware preparedness, 149
Hours of work
    flex time, 60
    maintenance, 206, 208
    overtime, 114
    staggered lunch hours, 60
    staggered maintenance and production
        crews, 173, 186
    staggered reporting schedules, 53
    see also Travel time
Human errors, 95, 98, 119, 131
Hygiene, 122, 130

IC. *See* Initial investment cost
IC (integrated circuits), 3
IE. *See* Industrial engineering
Incentive work environment, 10
Indexes, for productive management mea-
    surement, 299-300
Indirect-type quality components, 100-105
Industrial engineering
    cost consciousness, 259, 263
    equipment technology, 247-84, 285
    goal-setting, 263, 264
    know-how development, 12, 278-80

Industrial engineering (*continued*)
  for maintenance tasks, 86, 149, 196, 198
  and mistakes, 256
  organization of tasks, 281-84
  reliability/maintainability designs, 74-75,
    256-57
  and technological innovations, 7, 22
  see also Equipment engineering technology; Value engineering
Industrial engineers. *See* Equipment engineering technology; Industrial engineering
Inferior products. *See* Scrap
Information sheet, maintenance prevention,
  274, 277
Initial investment cost (IC), 265
Initial production flow management, 272-78,
  285
Initiative. *See* 5S activities; Motivation; Self-initiated maintenance
Innovation. *See* Technological innovation
In-operation inspection. *See* On-stream inspection
Input, interrelationship with output and management methods, 27-29
Inspection
  "cleaning is," 232
  cycles, 194, 199
  daily, 237-41
  diagnostic techniques, 189-201
  for equipment life-span, 49
  main concept of, 195-96
  on-stream, 172, 191, 194-95
  overseas techniques, 196
  patrol method, 196
  for preventive maintenance, 189-201, 237-41
  progress with PM evolution, 195-98
  purpose and value of, 189-92, 237
  selection of points of, 73
  self-initiated, 237-41
  skills, 241-44
  step reduction, 86-87
  table, 239-41
Insurance stocking method, 223
Integrated circuits. *See* IC
Inventory
  buffer, 114
  excessive spare parts, 223-24
  just-in-time system, 14-16, 113-14
  spare parts, 108, 215, 217-24
  zero WIP parts goal, 16
Investment
  in equipment and utilization rate, 46-48
  initial cost, 265
  see also Costs; Cost tree
Isolated repair work plan, 187-89

Japan
  division of labor in, 8-9
  energy conservation technology, 2-3
  equipment inspection techniques, 196
  equipment investment and utilization rate,
    46, 47
  5S concept, 122-33
  flexible manufacturing system, 5
  foreign-trade competitiveness, 4
  Gross National Product, 1, 2
  lifetime employment system, 10-11
  PM in petrochemical and steel industries,
    17-19
  "poka-yoke" (foolproofing devices), 92-93
  predictive maintenance techniques, 198,
    199
  quality level maintenance, 86-87
  and variety of plant management methods,
    23
Japanese Quality Control Award, 95
JIT. *See* Just-in-time system
Job-site training, 294-95
Just-in-time (JIT) system, automotive assembly, 14-16, 113-14

"Kanban" (task-direction note), 14, 15

Labor. *See* Work force
Labor unions, U.S., and work specificity, 8
Large-scale integration (LSI), 3-4
LCC. *See* Life cycle costing
Life cycle
  definition of, 249-51
  of equipment, 247, 248, 277
Life cycle costing (LCC), 22-23, 76, 248-49, 259, 262, 263, 265
Life span
  extension of equipment, 191
  inspection for, 49
  of parts, 73
  of spare parts, 226
Lifetime employment system, 10-11
Line supervisors, 301-9
Long-term breakdowns, countermeasures,
  213-14
Looseness prevention, 235
  see also Tightening of equipment
Losses, from equipment breakdowns, 55-59
LSI. *See* Large-scale integration
Lubrication maintenance activities, 156, 231,
  232-34, 237

Machine-Quality Analysis (M-Q Analysis
  Method), 89-91
  for equipment reliability improvement,
    162-63

Machine-Quality Analysis (M-Q Analysis
     Method) (*continued*)
   and QL relationship, 98-100, 103
Maintainability, 163-65, 201-14
   definition of, 201
   designs for, 74-75, 256-57
   improvement of, 205-8, 212-13
   and maintenance work characteristics,
      201-5
   measures against long-term breakdowns,
      213-14
Maintenance. *See* Inspection; Maintainabil-
     ity; Planned maintenance; Predictive
     maintenance; Preventive maintenance;
     Productive maintenance; Self-initiated
     maintenance; Total productive mainte-
     nance
Maintenance cards, 70-71, 87-88
Maintenance crew
   maintenance plan for, 169-70
   as maintenance skills training instructors,
      297
   sense of responsibility, 227
   separate from production department, 227
   skill integration, 186
   spare parts vouchers for, 225-26
   staggered hours, 53, 60, 173, 186
   systematization of tasks, 147-52, 186
   technical skills, 16, 86, 149, 151, 152,
      196, 198, 204, 227
   work activity elements, 209
   work during stoppages, 61, 156, 172-74,
      186
Maintenance department
   annual maintenance plan, 178-79
   calendar, 170-76, 185
   communication with, 210
   diagnostic techniques, 193
   failure analysis, 165-68
   independence of, 227
   methods, 176-78
   plan and breakdown-interval irregularity,
      205-6
   plan determining work quality, 169-70
   plan for isolated repairs, 187-89
   plans, 178-89
   role in effective equipment use, 286, 287
   and self-initiated maintenance, 227-46
   spare parts management, 215-26
   study of maintainability improvement,
      163-65
   systematization of tasks, 147-52
   system design, 152-57
   system efficiency, 154-55
   system function failure, 154
   technicians, *see* Maintenance crew, techni-
      cal skills
   transition in work activities, 227-29

Maintenance department (*continued*)
   troubleshooting manual, 163
   work characteristics, 201-5
   work manual, 211, 212
Maintenance prevention (MP)
   engineering technology design for, 262,
      263, 267
   in framework of preventive maintenance,
      247
   management of, 274-78
   see also Planned maintenance; Predictive
      maintenance; Preventive maintenance
Maintenance records, 70-85, 149
   calendar, 170-76, 185
   and daily inspection, 237, 239
   MTBF Analysis, 72-85, 109-10
   problems obtaining, 70-71
   values and goals, 71-72
Maintenance skills
   equipment-related, 191, 285-97
   technical, 149-52
   see also Maintenance crew
Maintenance work hours, 206, 208
   see also MTTR analysis
Management. *See* Equipment-oriented man-
     agement; Management training pro-
     grams; Managers; Plant management
Management training programs, 12
Managers
   attendance at TPM diagnostic/evaluation
      sessions, 41
   improving ability of line, 301-9
   motivation of small PM work groups by,
      37
   recognition of value of TPM activities, 21
Manual production lines, productivity-build-
     ing for, 51-54
Manuals
   for equipment use, 75-76, 236
   5S, 141-42
   for maintenance troubleshooting, 163
   of maintenance work activities, 211, 212
   for management by observation, 135
   for self-initiated maintenance, 243, 244
Manufacturing
   automation and, 4-5
   and equipment engineering technology,
      251
   multiprocess, 3
   see also Assembly; Factory; Flexible man-
      ufacturing system; Production
Mapping techniques, 76-78
Mean time between failures. *See* MTBF
     Analysis
Mean time to repair analysis. *See* MTTR
     Analysis
Mechanization, 1-3
Mentors, 294

Monthly maintenance plan, 179-87
  appropriation, 183
  calendar, 176, 185
  constraining factors, 183
  estimates in, 182
  flexibility, 183-86
  item organization, 181-83
  progress and evaluation, 186-87
  steps in creation, 181
Morale promotion, 126, 227
Motivation, 122-33, 155, 293
  see also 5S activities; Incentive work environment
MP. *See* Maintenance prevention
M-Q analysis. *See* Machine-quality analysis
MTBF Analysis, 72-85
  goals and application of, 73-76
  Table, 75, 76
  characteristics, 78-81
  creation, 76-85
  for equipment reliability improvement, 162-63
  for maintainability improvement, 163-64, 205
  for maintenance cost patterns, 109-10
  patterns, 81-84, 109-10, 156
  use of, 84-85, 163-64, 182, 262
MTTR Analysis, 205, 206, 208-9
  calculation of, 208
Multiple skill training programs, 287-92
Multiprocess manufacturing, 3

NC. *See* Numerical control
Nippon Steel, 19
Numerical control (NC), 3, 4

Objectives, of maintenance system, 153
Obsolescence, 10, 218, 223-24, 251
Oil crisis. *See* Energy conservation technology
Oils. *See* Lubrication maintenance activities
One-point lessons, 291-92
On-line inspection. *See* On-stream inspection
On-stream inspection (OSI), 172, 191, 194
On-stream repair (OSR), 172
Operational departments, and maintenance plan, 168-69, 172-73
Opportunistic maintenance, 173-74, 186
Order *(seiri)*, 122, 127-28
OSI. *See* On-stream inspection
OSR. *See* On-stream repair
Output
  efficiency improvement, 55-59
  efficiency and management of time-stamped per time period, 50-54
  evaluation of maintenance, 289-91

Output *(continued)*
  interrelationship with input and management methods, 27-29
  production line improvement measures, 53
  six factors of, 27
Overmaintenance, 198
Overtime work, 114

Pair method, 294
Pareto analysis technique, 87
Parts
  determining standards for, 73-74
  estimating life span, 73
  fabrication time, 211
  flow of, *see* Assembly
  optimized replacement cycle, 169
  replacement technology, 151-52
  standardization of, 225
  tightening of, 230, 231, 234-35
  zero WIP inventory goal, 16
  see also Spare parts
Patrol inspection method, 196
PC. *See* Production control
Periodic fixed-term maintenance, 176-77, 178
Personal cleanliness *(seiketsu)*, 122, 130
PERT. *See* Program Evaluation and Review Technique
Petrochemical industry, 15, 16-18
Planned maintenance, 165-89
  and predictive maintenance, 198-201
  and spare parts inventory, 224
  specific plans, 178-89
  see also Inspection; Maintenance prevention; Preventive maintenance
Planned production, 27-29
Planned purchase method, 222-23
Plant management
  analysis/diagnosis of standards, 43-145
  current standards, 43-44
  effective elements, 27-30
  and efficiency of equipment operations, 49
  and equipment engineering technology, 247-84
  and equipment repair cost-efficiency, 108-9
  and 5S activities, 32, 122-23, 131-45
  goals of, 29, 31, 32-34, 302-5
  interrelationship with input and output, 27-29
  leadership in skills training, 292-94
  and maintenance prevention system, 274-78
  by observation, 131-45, 156
  and overall productivity, 30-34
  and PM, 23-31, 300
  rally use by, 301-3

Plant management (*continued*)
    and spare parts policies, 215-16
    see also Equipment-oriented management;
        Managers
Plant security evaluation (PSE), 115-22
    implementation, 117-19
    reliability evaluation, 118-19
PM. *See* Preventive maintenance; Productive
    maintenance
"Poka-yoke." *See* Foolproofing devices
Policy-based life cycle, 249
Pollution. *See* Antipollution technology; Ex-
    haust restriction; Safety, environment,
    and pollution factors
Preautomation, 87
Precision, defects in, 231-32
Predictive maintenance, 177, 178
    concept of, 198-99
    diagnostic techniques, 189-92
    examples of philosophy promotion, 200-
        201
    goals of, 190, 199-200
    technical topics, 200
    techniques, 199
Preventive maintenance
    calendar, 170-73
    and continuous production, 15
    daily inspection, 237-42
    detailed plan, 167-68
    effectiveness measurement, 299-311
    equipment designed for, 267
    and equipment engineering technology,
        247-84
    equipment failure analysis, 165-68
    to extend equipment life span, 191
    inspection for, 189-201, 237-42
    and inventory turnover rate, 108
    item identification, 179, 180
    managerial aspects, 247
    maximum economical for operations and
        production needs, 168-69
    methods, 177-78
    opportunistic, 173-74
    and outdated equipment, 47
    plan determining work quality, 169-70
    planned repair program, 191-92
    plans, 178-87
    and production calendar, 172-73
    for quality, 88-89
    self-initiated, 123, 156, 227-29
    on shop floor, 230-37
    total employee participation in, 60-66
    in TPM practice, 167
    in workers' daily routine, 114
    see also Maintenance prevention; Planned
        maintenance; Predictive maintenance;
        Productive maintenance; Total produc-
        tive maintenance

Prioritized equipment, 62, 63, 168
Process capability (CP), 275, 276
Processing industry
    implementation of PM activities, 55
    production methods applied to mechanical
        assembly, 15-16
Procurement
    planned purchase method, 222-23
    of spare parts, 210, 211, 215
Product costs, 257-59
Production
    activities compared with maintenance ac-
        tivities, 201-4
    analysis of volume-equipment relationship,
        44-50
    cost and equipment quality, 105-13
    creating high productivity/zero breakdown
        lines, 44-54
    creating quality lines of, 93-106
    delivery date-equipment relationship, 113-
        14
    and division of labor, 227
    equipment design for, 269
    equipment as major means of, 1-2
    and equipment precision, 5
    initial flow management, 272-78, 285
    just-in-time system, 14-16, 113-14
    line accidents and stoppages, 119-22
    maintenance plan, 168-69, 172-73
    material flow management, 262
    planning and scheduling, 27-29
    processing industry methods applied to
        mechanical assembly, 15-16
    quality assurance line (QL), 95-105
    self-initiated maintenance, 228-37
    separate maintenance squad, 227
    start-up time, *see subhead* initial flow
        management
    work in progress, 14-16
    see also Assembly; Equipment; Safety, en-
        vironment, and pollution factors
        (SEP)
Production chart, 171-73
Production control (PC), 23
Production department, role in effective
    equipment use, 286, 287
Productive maintenance (PM)
    advancement system design for, 155-57
    analysis of current management standards,
        43-44
    as broad technology, 85-86
    cost issues, 27, 105-13
    and departmental goals implementation,
        31-34
    diagnosis/evaluation/implementation sys-
        tem, 38-41
    effects, measurement of, 310-11
    and equipment efficiency, 44-59

Productive maintenance (PM) (*continued*)
  and equipment engineering technology,
    253
  and equipment inspection activities, 195-
    98
  evaluation of, 38-41, 299-311
  and 5S activities, 122-33
  and group activities, 37-38
  importance of theme selection, 37-38
  improvement plan, 157-62
  management commitment to, 37-38
  MTBF Analysis for, 72-85, 109-10
  and output efficiency improvement, 55-59,
    85-86
  in petrochemical and steel industries, 16-
    19
  planned and chance, 165-89
  planning and managing, 147-226
  and plant management methods, 23-31
  plant security system relationship, 119-22
  and production demands, 16
  and productivity, 44-54, 291
  rally, 301-3, 305-8, 309
  realities of ongoing, 43-44
  record-keeping and use, 70-85
  self-initiated, 227-46
  skills and skills training, 13, 285-91
  and spare parts inventory, 217, 218
  study of maintainability improvement,
    163-65
  synchronized plan by batches, 48
  system design, 152-57
  system and tasks, 147-52
  technical skills for, 16, 86, 149, 151, 152,
    196, 198, 204, 227
  and technological innovations, 2-5
  worker involvement in, 7-8, 11-13
  and zero breakdown, 44-50
  and zero defect, 11-12
  and zero maintenance measure, 162
  see also Maintenance prevention; Preven-
    tive maintenance; Predictive mainte-
    nance; Total productive maintenance
Productivity
  and breakdown improvement, 57
  improvement, 50-54
  PM-building to increase, 44-54, 291
  varied management methods to attain, 30-
    34
Product quality
  definition of, 289
  see also Quality control
Program Evaluation and Review Technique
    (PERT), 187-89
Programming technology, 3
Promotional activities
  for companywide TPM advancement, 34-
    38

Promotional activities (*continued*)
  PM rally, 301-3, 305-8, 309
  see also 5S activities
PSE. *See* Plant security evaluation
Purchases. *See* Procurement

QC. *See* Quality control
QL. *See* Quality assurance production line
Quality
  components management, 100-105
  equipment and production cost, 105
  factors affecting equipment, 231-32
  maintenance, skills guaranteeing, 289-91
  of repairs, 204-5, 289
Quality assurance production line (QL), 95-
    105
Quality control (QC)
  automated, 4-5
  equipment relationship, 75, 85-105
  and 5S activities, 125-26
  foolproofing devices, 92-93
  goals and values of maintenance activities,
    86, 162
  of maintenance work, 169-70
  M-Q analysis, 89-91
  Pareto analysis technique, 87
  perfect production lines and, 93-106
  and PM activities, 11-12
  preventive maintenance for, 88-89
  as product selling point, 253
  and technical skill training, 12
  three principles of, 86
  zero defect, 11-12, 93, 289

Rally, PM, 301-3, 305-8, 309
Realities, of ongoing PM, 43-44
Record keeping
  daily inspection, 237, 239
  foolproofing logs, 93
  maintenance, 70-85, 149, 173-74, 185
  MTBF Analysis, 72-85, 109-10
Recycling, of spare parts, 218, 223, 225
Reliability
  design checkpoints, 271
  designs for, 74-75, 256-57
  equipment, 74-75
  equipment improvement, 157-64
Renovated equipment, 47
Repairs
  actual time analysis, 209-12
  adjustment and tryout time, 212
  block replacement method, 216
  cost-efficient management system for,
    108-9
  cost of, 107
  cost reduction activities, 110-13
  diagnostic time, 210
  disassembly time, 211

Repairs (*continued*)
  emergency technical skills, 151
  during equipment operation, 194
  and equipment life cycle cost, 249
  improving methods, 74
  isolated work plan, 187-89
  maintenance crew for, 147, 149
  MTTR analysis, 205, 206, 208-9
  notation on MTBF Analysis Table, 85
  optimized plan, 73
  parts procurement time, 210
  planned program for preventive mainte-
    nance, 191-92
  quality of, 204-5, 289
  skill-integrated maintenance department,
    186
  and spare parts inventory, 107-8
  time standards, 74
  and travel time, 209-10
Replacement, and equipment life cycle costs,
    249, 250-51
Responsibility
  concept of clearly defined individual, 8-9,
    10
  for establishing operational standards and
    maintenance work, 74
  maintenance crew's sense of, 227
  for TPM, 303, 305
  see also 5S activities
"Road construction method" maintenance,
    168
Robots, 3, 4, 30-31
Rules, 131, 133

Safety, environment, and pollution factors
    (SEP)
  equipment relationship with, 115, 277
  and 5S activities, 122-45
  plant security evaluation and, 115-31
SC. *See* Sustenance costs
Scheduling
  production, 27-29
  see also Calendar, maintenance; Planned
    maintenance
Scrap
  and equipment precision, 5
  and inspection methods, 87
  reduction and equipment technologies, 86
  and setup time, 58
Security systems. *See* Plant security evalua-
    tion
Seiketsu (personal cleanliness), 122, 130
Seiri (order), 122, 127-28
Seiso (cleaning), 122, 129-30, 131, 230,
    231-32
Seiton (tidiness), 122, 128-29
Self-initiated maintenance, 227-46

Self-initiated maintenance (*continued*)
  activities, 123, 155, 156-57
  background of, 227-29
  Certification System, 246
  high priority activities, 229-37
  problems to check, 231-37
  skill training for, 242-44
  steps toward, 244-46
  tools for, 241-42
  see also 5S activities
Seniority, 13
SEP. *See* Safety, environment, and pollution
    factors (SEP)
Setup time, and scrap, 58
Shitsuke (discipline), 122, 130, 131
Short stoppages
  analysis of, 67-70
  causes of, 68, 232
  remedies for, 69-70, 232
Simple fixed-quantity ordering methods, 220
Skills
  diverse and multiple, 30, 287-92
  equipment-related, 191, 285-97
  guaranteeing maintenance quality, 289-91
  inventory list, 12-13, 287
  raising level of, 288
  self-initiated maintenance training, 242-44
  training programs, 12-13, 30, 191, 292-95
  verification of, 291-92
  and workers' TPM involvement, 8-13
  see also Technical skills
Slogans, 95, 124
Spare parts
  breakdown patterns, 277-78
  inventory characteristics, 217-18
  inventory management method, 16, 219-
    23
  life span extension, 226
  for maintenance as differentiated from pro-
    duction, 217
  management of, 215-26
  obsolescence, 218, 223-24
  outside orders vs. in-house overhauling,
    225
  procurement, 211, 215
  recycling, 218, 223, 226
  replenishing, 219
  review in maintenance prevention design,
    271-72
  stocking methods, 218
  storage location, 218-19
  storage and repair time, 107-8
  vouchers, 225-26
  see also Parts
Specialization
  broadening scope of, 10-11
  worker, 8-9

Specialization (*continued*)
   see also Division of labor
Staggered work schedules, 53, 60, 173, 186
Standardization
   of maintenance tasks, 206, 212
   of parts, 225
Standards, analysis/diagnosis of management, 43-145
Start-up time. *See* Initial production flow
   management
Steel industry, 19
Stoppages
   analysis of short assembly-line, 67-70
   block replacement method to reduce, 215-
    16
   brief, factors in, 67-70, 232
   equipment failure prevention, 158-62
   isolated repair work plan, 187-89
   line, causes of, 119-22
   maintenance activities during, 61, 156,
    172-73
   and opportunistic maintenance, 173-74,
    186
   see also Breakdowns
Storage, spare parts, 107-8, 218-19
Stresses and strengths, equipment, 192-94
Supervisors. *See* Line supervisors; Managers
Sustenance costs (SC), 265
   see also Life cycle costing
Symptoms, of breakdowns, 194-95

Tasks
   area reorganization, 12-13
   employee PM involvement, 17-19, 163
   just-in-time system, 14-15
   maintenance crew, 209
   organization of engineering, 281-84
   see also 5S activities; Self-initiated main-
    tenance
Technical skills
   for maintenance technicians, 16, 86, 149,
    151, 152, 196, 198, 204, 227
   for PM, 12-13, 149-52
   see also Equipment engineering technol-
    ogy; Maintenance skills; Training within
    industry
Technological innovation
   in 1980s, 2-5
   energy-conserving, 2-3
   Japanese, 1-3
   in slow-growth periods, 7
   and utilization of workers' skills, 10, 13
   see also Equipment engineering technol-
    ogy
Temperature control, 235-36, 237
Temporary purchase items, 223

Testing, in skill training, 294
Tidiness *(seiton)*, 122, 128-29
Tightening of equipment, 234-35, 237
Time
   analysis of repair, 209-13
   frame for MTBF Analysis Table data, 78
   standards for repairs, 74
   unclear usage, 58
   see also Hours of work
Time-stamped output, 50-54
Tools. *See* Equipment
Torque wrench, 235
Total productive maintenance (TPM)
   aims of, 19
   analysis/diagnosis of management stan-
    dards for, 43-145
   awards, 95
   basic plan and implementation, 21-41
   concepts effective for industries, 7-8
   corporate-wide advancement of, 7, 34-38
   diagnosis/evaluation/implementation sys-
    tem, 38-41
   goals, 31-34
   and preventive maintenance plans, 167
   and productivity improvement, 30-31
   requirements, 1-5, 21-22
   sphere of activities, 22
   staggered hours for maintenance crew, 53,
    60
   and workers' skill-development, 8-13
   work force responsibility for, 7-8, 11-13,
    16, 22, 30, 95, 305
   see also Maintenance prevention; Planned
    maintenance; Predictive maintenance;
    Preventive maintenance; Productive
    maintenance
Toyoda, Kiichiro, 14, 15
Toyota Motors, 5, 14-15, 113
TPM. *See* Total productive maintenance
"TPM to Enhance Quality and Productivity"
   (slogan), 95
Training within industry (TWI)
   equipment use skills, 191, 285-97
   5S activities, 131
   handcrafted program, 295-97
   job-site, 294-95
   for maintenance technological skills, 152
   management programs, 12
   for multiple skills, 287-92
   pair method, 294
   and skill promotion, 12-13, 30, 131, 191,
    292-95
   and skills for self-initiated maintenance,
    242-44
   skill verification, 291-92
   for technical skills, 12, 131
   testing in, 294

Travel time, maintenance crew, 209-10
TWI. *See* Training within industry

Unclear usage times, 58
United States, 8-9, 22-23
Utilities, 100

Value engineering (VE), 27
VCR (video cassette recorder), 3
VE. *See* Value engineering
Vibrometer measurement, 196
Video cassette recorder. *See* VCR
Volume, relationship analysis between equip-
    ment and production, 44-50
Vouchers, 225-26

Warehousing methods, 217
Weekly maintenance plan. *See* Monthly
    maintenance plan
WIP. *See* Work in progress
Work environment. *See* Factory; 5S activi-
    ties; Safety, environment, and pollution
    factors (SEP)
Work force
    and breakdowns prevention, 60-66
    and brief stoppages, 68
    correct operation of machinery, 236-37
    division of labor in, 8-9, 227
    division of labor and skill inventory, 12-
        13
    division of maintenance assignments, 73
    and efficient maintenance system, 155
    equipment use training, 191, 285-97
    and equipment utilization rate, 46
    familiarity with equipment, 242-44
    and 5S activities, 32, 122-45, 156, 305
    and foolproofing devices, 92-93
    goal management plan table, 303-5
    human errors, 95, 98, 119, 131
    importance of seasoned, 30

Work force (*continued*)
    Japanese lifetime employemnt system, 10-
        11
    and job-assignment scope, 9
    job-site practical training, 294-95
    maintenance operations costs, 107
    multiple skills training programs, 287-92
    overtime drawbacks, 114
    pair method, 294
    and PM group activities, 37-38
    and PM involvement, 12-13
    and PM tasks, 17-19, 163
    and preventive maintenance of tools, 230-
        37
    and raw maintenance data, 71
    and self-initiated maintenance, 227-46
    skill diversity, 30
    skilled maintenance personnel, *see* Mainte-
        nance crew
    skill-training programs, *see* Training
        within industry
    staggered maintenance and production
        hours, 173
    staggered work-reporting schedules and
        productivity, 53
    and TPM commitment, 7-8, 11-13, 16,
        22, 30, 95, 305
    see also Hours of work; Maintenance crew
Working hours. *See* Hours of work
Work in progress (WIP)
    inventory and just-in-time system, 14-16,
        113-14
    zero inventory goal, 16

ZD. *See* Zero defect
Zero breakdown
    equipment-quality map for, 93-94
    PM-building and, 44-50
Zero defect (ZD), 11-12, 93, 289
Zero maintenance measure, 162
Zero WIP parts, 16